Library of Congress Cataloging-in-Publication Data

ISBN 0-7923-1983-4

Published by Kluwer Academic Publishers,
P.O. Box 17, 3300 AA Dordrecht, The Netherlands.

Kluwer Academic Publishers incorporates
the publishing programmes of
D. Reidel, Martinus Nijhoff, Dr W. Junk and MTP Press.

Sold and distributed in the U.S.A. and Canada
by Kluwer Academic Publishers,
101 Philip Drive, Norwell, MA 02061, U.S.A.

In all other countries, sold and distributed
by Kluwer Academic Publishers Group,
P.O. Box 322, 3300 AH Dordrecht, The Netherlands.

Printed on acid-free paper

Printed in Belgium

Library of Congress Cataloging-in-Publication Data

ISBN 0-7923-1983-4

Published by Kluwer Academic Publishers,
P.O. Box 17, 3300 AA Dordrecht, The Netherlands.

Kluwer Academic Publishers incorporates
the publishing programmes of
D. Reidel, Martinus Nijhoff, Dr W. Junk and MTP Press.

Sold and distributed in the U.S.A. and Canada
by Kluwer Academic Publishers,
101 Philip Drive, Norwell, MA 02061, U.S.A.

In all other countries, sold and distributed
by Kluwer Academic Publishers Group,
P.O. Box 322, 3300 AH Dordrecht, The Netherlands.

Printed on acid-free paper

Printed in Belgium

NEXT GENERATION INFRARED SPACE OBSERVATORY

Edited by

S. J. BELL BURNELL, J. K. DAVIES
& R. S. STOBIE

*Royal Observatory, Blackford Hill,
Edinburgh EH9 3HJ, Scotland, U.K.*

Reprinted from Space Science Reviews, Volume 61, Nos. 1–2, 1992

KLUWER ACADEMIC PUBLISHERS
DORDRECHT / BOSTON / LONDON

Participants of the Workshop on the Next Generation Infrared Space Observatory

EDINBURGH 22–24 MAY 1991

1. B. Baud
2. P. Saraceno
3. Roger Emery
4. S. Fonti
5. Paul Harvey
6. Michael Rowan-Robinson
7. Harley Thronson
8. Colin Humphries
9. Simon Chase
10. Phil James
11. Alan Penny
12. Ray Wolstencroft
13. Steve Rothero
14. Paolo Persi
15. Ian Furness
16. Herman van de Stadt
17. Masa Hayashi
18. Ed Erickson
19. Tim Hawarden
20. Anna Orlowska
21. Eli Atad
22. Jon Davies
23. Ytrak Nevo
24. Ferrari-Toniolo
25. Vincenzo Natale
26. Rodger Thompson
27. Martin Wells
28. Rens Waters
29. Helen Walker
30. Selamet Köse
31. Jocelyn Bell Burnell
32. Bob Stobie
33. Martin Kessler
34. A. F. J. Morgovwnck
35. Richard Tuffs
36. Peter Brand
37. Steve Russell
38. Diah Anggraeni
39. Therese Encrenaz
40. Malcolm Longair
41. Maureen McLean
42. Chris Baddiley
43. Andy Longmore
44. Tom Broadhurst
45. John Davies
46. John Harris
47. Paul Murdin

TABLE OF CONTENTS

FOREWORD

The workshop on the Next Generation Infrared Space Observatory was hosted by the Royal Observatory, Edinburgh and took place at the Heriot Watt conference centre on 22-24 May 1991. Scientists, engineers and industrialists from Europe, the USA and Japan participated in two days of open and enjoyable discussion on possible strategies for space-based infrared astronomy in the next century. Many of these contributions appear in this issue as papers prepared on the basis of the talks presented at this workshop.

Where possible within the editorial constraints, the order of the papers follows that of the meeting, reviewing scientific issues, existing and planned space projects and then considering technical issues which may determine the nature of the next generation of missions. Although many of these issues remain open, it is hoped that these proceedings will act as a starting point, and as a spur, to consideration of how innovative technologies can be used to provide new opportunites for astronomers of the future.

Finally we would like to thank Maureen McLean, Dorothy Skedd and Eve Thomson of the Royal Observatory who helped to organise the workshop and to prepare the files which were used to compile these proceedings.

S J Bell Burnell, J K Davies, and R S Stobie
Guest Editors

WELCOME & INTRODUCTION

PAUL MURDIN
Royal Observatory, Blackford Hill, Edinburgh EH9 3HJ

(Received 22 May, 1991)

I've just learnt that the bar is open to 11 o'clock so I am going to be brief and perhaps Malcolm Longair, whom I am to introduce, will take the hint.

It's my pleasure to welcome you to Edinburgh – a new city for me to live in. I hope you get a chance to enjoy some of the attractions that I have been discovering here myself over the last few months. People tell me that this is in many ways the most attractive season in Edinburgh and perhaps its most attractive month of the year so enjoy your visit to our city.

This conference is set up to explore a new idea; this is appropriate in this setting because Scotland has a distinguished record in the generation of new ideas and new technologies. James Watt, Lord Kelvin and James Clerk Maxwell are Scottish scientists whose very names resonate with our meeting. I learnt today that James Dewar too was Scottish born. I should have realised this because I have often drunk a product which is internationally identified with this country and and labelled with his family's name.

Dewar started his career in the sciences at Edinburgh University; then he moved south and began to work on low temperature physics, his achievements in this area being recognised by the eponymous cryogenic vessel. When he left Edinburgh he became the then Jacksonian Professor at Cambridge; this is a path which has been re-trodden by our guest speaker here tonight.

On a local scale, ROE is proud of its own achievements in new ideas and new technologies. It operates innovative telescopes like UKIRT, with its lightweight mirror – a revolution in its day – and the James Clerk Maxwell Telescope, the world's largest submillimetre telescope with the most accurate surface. ROE builds innovative instruments for these telescopes – our infrared array camera, IRCAM, a newly commissioned grating spectrometer which we call CGS4, and a submillimetre array camera called SCUBA are recent and current examples. So we are very pleased to be hosting this meeting to develop the innovative idea we are all here to discuss, namely the Edison proposal.

The conference title is "The Next Generation Infrared Space Observatory". It seems like only yesterday when astronomical satellites were small instruments, run by dedicated project teams; if you were lucky you could get on the periphery of these insider teams. Now we are in the midst of the present generation of the great space observatories, available to a broad community of astronomers who can work in the middle of the missions – I think of the Einstein Observatory in the past, the

Space Science Reviews **61:** 3–4, 1992.

Hubble Space Telescope in the present, and the Gamma-Ray Observatory to come.

Edison is the logical development of a series of infrared space missions. IRAS is old and is no longer cold but the IRAS archive and its results are still living. We use IRAS results in the observatory on Mauna Kea on a day-to-day basis, perhaps hour-to-hour even.

ISO is yet to be born. We played a part at ROE in the optics installed on ISO, we helped at the birth, we will help its operation from Madrid and we look forward to the science that is yet to come.

SIRTF too is yet to come – in gestation.

Clearly we need to conceive the infrared mission beyond this series. The infrared missions seem to be quantised in decades. IRAS was 1983, ISO will be 1993, SIRTF perhaps 2003? Extrapolating my series in decades, Edison will be the mission of 2013.

Edison's unique selling point is the passive cooling. I have always been surprised that it has been possible to launch successfully a satellite containing a tank of liquid helium. The lifetime limitation is in any case a serious constraint which restricts and drives and constrains the scientific programme of such satellites. Passive cooling in Edison will be the key to free infrared astronomy from this constraint, but we need to discuss the potential problems, the likely developments and other factors that will be the subject of this workshop here this week. Incidentally, this is a workshop – don't feel inhibited about contributing ideas to the discussion.

Thomas Edison, who unfortunately for the literary style of this introduction was not a Scotsman, shed light all over the world with his invention and perhaps the satellite which we expect to be named after him, Edison, will also spread light in the early twenty-first century throughout the world of astronomy. We need to identify the scientific programmes on which the light can be expected to be shed; the scientific goals and the key technologies will interact in this next infrared observatory.

We recognise that Edison will be expensive and that international collaboration is going to be essential. We welcome especially the guests to the workshop from overseas. Thank you for making the extra effort to come and be with us this week. I look forward to seeing you again at Cape Canaveral at the launch of Edison in 2013, maybe even a little sooner.

I would like to introduce to you our guest speaker, Malcolm Longair, my very distinguished predecessor as Director of ROE and now Jacksonian Professor at the University of Cambridge, to talk to us about what lies beyond 2013, "The Promise of Infra-Red Space Astronomy".

THE FUTURE OF SPACE INFRARED ASTRONOMY

M.S. LONGAIR
Cavendish Laboratory, Madingley Road, Cambridge CB3 0HE

(Received 20 June, 1991)

1. Foreword

I have been asked to make some opening remarks about the future of space infrared astronomy, with special reference to the very exciting new ideas which are the subject of this workshop. I take literally the intention of the organisers that this is a "workshop" and so everything which the participants and I say are really ideas, concepts and work in progress. This is particularly important for the concept of passively cooled space infrared telescopes where there are many open issues. My role is to be provocative so that we at least address as many of the important questions as we can without necessarily being able to answer them all.

Let me begin by defining the terms POIROT and EDISON. I use term a POIROT to be any **P**assively-cooled, **O**rbiting, **I**nfra-**R**ed, **O**bservatory **T**elescope. This was the name of the proposal which Tim Hawarden presented to the European Space Agency's Astrophysics Working Group meeting in 1989. No one will persuade me that the splendid acronym was not influenced by the fact that the chairman of the Working Group was Belgian! I will use the term to refer to all classes of passively cooled space infrared telescope, from diameters we are used to, say 60 cm, to the large infrared telescope of one's dreams, 8-metres or larger, which is perhaps not so much a fantasy as it was a few years ago. I will use the term EDISON to refer to the specific proposal developed by Harley Thronson, Tim Hawarden and a host of others which may be thought of as a concrete realisation of the generic class of POIROTs.

2. The Status of Infrared Astronomy

No one at this workshop needs any reminding of the enormous revolution in infrared astronomy which has taken place over the last ten years. The technology of infrared detector arrays and advances in all related aspects of instrumentation – mechanical, electronic, optical and cryogenic design – have led to quite staggering advances in scientific capability. Most major observatories now have programmes of advanced infrared instrumentation as a matter of course – here, in Edinburgh, the successful commissioning of the two-dimensional infrared array spectrograph, CGS-4, is a beautiful example of what can now be achieved. What lies in the future

Space Science Reviews **61:** 5–12, 1992.

is illustrated by the quality of the detector arrays being prepared for the NICMOS instrument which will fly on the Hubble Space Telescope. All of us here know that these advances will have the most profound impact upon the whole of astrophysics.

I want to make only two comments on the science, both being exhortations to the community to enhance awareness of these advances among the wider astronomical community and to increase the constituency who should be capitalising upon them. The first point relates to the fact that it is still mostly the pioneers and afficionados who are using these wonderful instruments. In other words, most of the real excitement is still within the club of those who consider themselves to be primarily infrared astronomers. From my perspective, the advances are so fundamental that all astronomers should be looking automatically to the infrared wavebands to advance their scientific programmes. As a long time convert, I am still powerfully impressed by how much simpler many things become in the $1-5\ \mu$m waveband as compared with the optical waveband. Let me cite a few examples, not far from my own interests – everyone could make up their own list. For example, $2\ \mu$m observations are clearly to be preferred in such important correlation studies as the period-magnitude relation for R-R Lyrae stars, the infrared Tully-Fisher relation and the infrared Hubble diagram for the brightest galaxies in clusters and radio galaxies.

It is important to note that there are two good reasons why the infrared observations are to be preferred. The first is the obvious one that the problem of dust obscuration can essentially be eliminated. The second is more physical and is related to the fact that, in general one is working in the Rayleigh-Jeans region of the spectrum for most stars in the infrared and that region of the spectrum is much more stable than the Wien region in which the intensity decreases exponentially and absorption lines can strongly influence the energy distribution. This is a well-known story but it is one which needs to be repeated over and over again. The point of this argument is that, when the case is being made for a major new astronomical facility, such as a POIROT, it is plainly of the greatest importance that the scientific case for the facility is very broadly supported within the whole astronomical community and not just among the infrared astronomers.

The second point concerns the science product of the new generation of infrared instruments. We all know of the wealth of data which has already been taken by these instruments but I personally would like to see much more of the material in the scientific literature. There is, of course, a time-lag between new facilities becoming available and the appearance of a substantial body of scientific achievements in the astronomical journals. I firmly believe, however, that only when these major contributions are recognised by everyone, will the subject have grown to real maturity. This paragraph therefore amounts to a plea to all those with new infrared results to ensure that these are rapidly published and their full import appreciated by the whole community. It will be essential to further the case for the next generation infrared telescope that the scientific case is absolutely first-class.

There is no need for me to list here the scientific projects which form the backbone of infrared studies – they span the whole of astronomy and cosmology and, in many cases, the science can be uniquely addressed in these wavebands. These are well summarised in Harley Thronson's manifesto – brown drawfs of all sorts, star formation with high angular and spectral resolution, dust-free studies of active galactic nuclei, galaxies at large redshifts, protogalaxies. It is a familiar list but the problems have become no less pressing as time goes on. I believe the case for all these studies will become stronger and stronger as the problems are refined by ground-based and space observations which will be made before the next generation of space infrared telescopes flies.

3. The Next Generation Space Infrared Observatory

The nature of the next generation space infrared telescope should be looked at in the context of the future missions which are scheduled to fly – the Infrared Space Observatory (ISO) of ESA and the Space Infrared Telescope Facility (SIRTF) of NASA. I would claim that the 2 – 100 μm wavebands are still relatively unexplored and that ISO and SIRTF are primarily *discovery missions*. I say this because of the limited lifetimes of these missions and the undoubted problem of planning a major astrophysical campaign when we do not really know what the key problems or discoveries wil! be. I am therefore very firmly of the view that it is correct to consider the Poirots to be the generation of large space infrared telescopes beyond the epoch of ISO and SIRTF. The corollary of my argument is that space infrared astronomy will need an observatory class facility which will have a long operational lifetime in space.

The key characteristic of the next generation telescopes will be *large aperture*. This provides simultaneously high angular resolving power as well as much enhanced sensitivity for all types of observation. Indeed, we might use this argument backwards and state that, because it is a prerequisite to obtain large aperture, the concept of a passively-cooled infrared telescope is almost certainly the only practical way of obtaining at least an order of magnitude improvement in resolution and sensitivity for infrared astronomy. I believe that the case can be made that it is not a question of *whether or not* passively cooled telescopes are a desirable option but rather that they are likely to be the *only* way of building the next generation large space infrared telescopes.

Let us explore the question of the optimum size of a POIROT. I would argue that the minimum size we should be considering for an observatory-class mission is 2.5 m, essentially the EDISON mission as described by Harley Thronson. It is also intriguing to discuss what the maximum size could be. I have in mind the statements made at the Goddard meeting on *Observatories in Earth Orbit and Beyond* that the Soviet Energia launch vehicle is capable of launching a payload of 100 tonnes into Earth-orbit. At the Baltimore IAU meeting, Kardashev described the feasibility of launching 8-metre optical telescopes with such a launch vehicle. Therefore, unlike the situation a few years ago, there is no reason in principle, why we should not consider telescope apertures in the range 2.5 to 10 metres.

This, therefore, leads to my first questions to the technologists – *what is the cost-diameter relation for POIROTs?* It is not at all obvious that it is of the same form as that of ground-based telescopes which are dominated by the size of the telescope enclosure and the increasing mass of the structure to prevent the telescope bending under gravity. In fact, in discussion with Rodger Thompson, he stated that the only known relation between the cost of a space project and its size is that the cost is roughly proportional to the number of technical drawings which have to be made! This reflects the fact that, in the end, the cost is largely driven by the manpower effort working on the project. There is an interesting lesson here for the design of a large space infrared telescope, if Rodger's rule is indeed true. It is of the greatest importance to have much more accurate estimates of the cost-diameter relation for POIROTs.

The next question is one of pointing accuracy. I assume that grinding and polishing the mirror to a good figure is not a problem. I say this because I assume that the shortest wavelength at which the telescope should be used is 2 μm and therefore we are not seeking optical quality performance. However, the pointing accuracy is very important. To a good approximation, the diffraction limit, $\theta = \lambda/D$, is $0.4/D$ arcsec at 2 μm where D is the diameter of the primary mirror in metres. We have to recall that the pointing accuracy has to be about ten times better than the angular resolution or else the diffraction limited images are degraded. Therefore, we see that, if the telescope is to be diffraction limited at 2 μm, the fine guidance system has to be able to maintain the telescope pointing in the same direction with an accuracy of 0.01 arcsec, i.e. we are in the same ball-park as the Hubble Space Telescope! None of us forgets the problems there have been in achieving this pointing accuracy. In fact, this problem can be solved by changing the approach to the fine-guidance problem by using large format CCDs rather than the beam splitting techniques which have been employed in the HST.

All my instincts are to go for the largest POIROT which can be built. I would regard the EDISON concept as being the minimum scale project for the generation of instruments after ISO and SIRTF. In fact, if we look at the deliberations of the working groups on the Next Generation Large Optical Space Telescope, we find that they are thinking along remarkably similiar lines. The gains in scientific capability with telescope diameter are so great that I believe we should consider the next generation infrared telescope as being the infrared counterpart of the next generation large space optical telescope. The time-scale over which one might hope to launch such a telescope is similar to that which is being talked about by the proponents of the next generation large optical space telescope. My guess is that, at the earliest, we are talking about a launch in the latter half of the first decade of the next century

4. Technical Goals

4.1. THE TEMPERATURE OF A POIROT

The key questions here are how cool the telescope and focal plane structures can become with optimised passive cooling. I am sure that Tim Hawarden will report on the work which has been carried out to model the expected thermal behaviour of POIROTs of different sizes. We also need to know how sensitive to temperature the different parts of the telescope will be when it points in different directions.

We also need to know how much cooling power we can hope to generate by the use of closed-cycle coolers for cooling the detectors in the scientific instruments. The whole question of thermal stability and which parts of the telescope may need to be heated or cooled are important questions.

4.2. INSTRUMENTATION

The planning for the instrumentation should run in parallel with the design proposals for the telescope itself. Since this is a programme for the next century rather than the present one, we have to guess what technology will be available over that time-frame. My guess is that we need to set goals for very large infrared arrays, say 1000×1000 pixel arrays, for all the wavebands between 2 and 100 μm. The spectrographs will undoubtedly be two-dimensional systems like CGS-4 but fully space qualified and with very large arrays.

There is every advantage in these development programmes beginning well ahead of the date by which one might expect the project to be approved. It is worthwhile recalling that the CCD option for the Wide Field/Planetary Camera of the Hubble Space Telescope was adopted before the CCDs had really been used astronomically on ground-based telescopes. The camera is now a great success but many important lessons about the performance of CCDs had to be learned while the instrument was being built. It is clearly advantageous if that situtation can be avoided.

In this planning, the instrumentation should be designed as an integral part of the telescope project from the very beginning. Because the instruments will almost certainly have to be cooled, their design cannot be separated from that of the spacecraft itself.

4.3. ORBITS

There are a number of compromises which have to be made in selecting the optimum orbit for a POIROT. Easy accessibility means low Earth orbit but almost certainly we would not want to deal with the thermal problems which would result from the bright Earth fully illuminating and heating the telescope every orbit. We therefore need to consider either very high orbits, long elliptical orbits, such as that chosen for ISO, or possibly locating the observatory at the outer Lagrangian point of the Earth-Sun system. This last possibility has the great attraction that the Sun and Earth remain in roughly the same position relative to the spacecraft and thus minimises the thermal variations. I hope this possibility will be considered in some detail.

4.4. Optical Design

There is a need to carry out optical design for the complete telescope system including the scientific instruments. Colin Humphries and his team will be reporting on preliminary studies for the optical design of POIROT and EDISON.

4.5. Simplicity versus Complexity

One of the vexed topics in designing any observatory-class instrument is the question of complexity versus simplicity. It is generally true that the most successful space missions to date have been what might be called "simple" missions by which I mean that they have been largely single purpose instruments. I am thinking of missions such as IUE, COS-B, IRAS and COBE. Complexity necessarily brings with it greater risk and the Hubble Space Telescope is a good example of the risks involved if a complex observatory-class mission is built. Let me hasten to add that I believe the HST will be a great success. It is a tragedy that the programme has been so severely impacted by the problem of spherical aberration of the primary mirror. But for that, the project would already be hailed as a great success.

Opinions differ about how great a question the issue of simplicity versus complexity is. In terms of cost, I have a strong feeling that Rodger Thompson's rule that cost is correlated with the number of technical drawings contains a basic truth and this would argue in favour of simplicity. However, there will always be strong temptations to increase the complexity of any large telescope to provide enhanced scientific capabilities. This issue will have to be fought out at the mission definition stage.

4.6. Costs

Robert Smith's excellent book *The Space Telescope* is essential reading for anyone with an interest in space astronomy. The origins of the cost overruns in the programme were inherent from the beginning. I remember that, at the very first meeting of the HST Science Working Group, the project was described as a "low-cost, success-oriented programme". We might be able to live with the term "low-cost" but to term a programme as complex as the HST "success-oriented" was plainly asking for trouble, as subsequent events showed. We really need to persuade the powers that be that we really know what this programme would cost. If the diameter of the telescope is 2.5 metres or greater, I believe we are considering a mission in the "Great Observatories'" bracket, both scientifically and financially, and rightly so. We need to know these costs with some precision. Please put Robert Smith's book on your reading list.

4.7. MANAGEMENT

In my view, the next generation infrared space telescope is Big Science and it seems inevitable that we have to call upon the national and international agencies to cooperate in making it feasible. Management is the key to success in missions like this and so we have to use expertise of these agencies to provide the management infrarastructure for the project. University and Observatory groups have the key role of driving the science, innovative technology and instrumentation for the mission.

4.8. MINI-POIROTS

One gap in my vision of the future of space infrared astronomy is the need for a deep infrared survey of the sky. This project would fulfil the role which the deep Schmidt telescope surveys perform for optical astronomy and which the ROSAT sky survey will perform for X-ray astronomy. It is highly desirable that such a deep infrared survey is carried out before the next generation infrared space telescope is launched. One very attactive way of doing this is to construct a mini-POIROT which would be a small scale version of a mission such as EDISON or super-POIROT. Besides the sky survey, the mission would provide essential information about the performance of real POIROT-type missions. Since it would be a dedicated mission, not dissimilar from IRAS, it could be built much more cheaply than EDISON and could well be accommodated within the space programme of a single country. The spectacular results obtained from the IRAS survey show how much frontier science can come from such a mission.

5. Conclusion

I can remember vividly when Tim Hawarden first came into my office about 7 years ago and talked very excitedly about passively-cooled space infrared telescopes. He was sufficiently persuasive that we even managed to get the idea into the various forward-look exercises in which the SERC indulges every year. The idea has evolved considerably since then and the climax of that activity was when Tim presented the case for POIROT to the European Space Agency a couple of years ago. The project did very well indeed as it passed through the peer-review committees and was unlucky just to miss being selected for further study.

Since then, the momentum has continued largely through the enthusiasm of Tim and of Harley Thronson's advocacy in the USA. It is my view that the time has come when the agencies have to take these proposals very seriously indeed as the way forward for space infrared astronomy. Our goal over the next two days is to expose all the issues which have to be addressed to convert the concept into a realistic space mission.

References

Kondo, Y. (ed): 1991, *Observatories in Earth Orbit and Beyond*, Kluwer Academic Publishers, Dordrecht

Smith, Robert W.: 1989, *The Space Telescope*, Cambridge University Press, Cambridge

INFRARED SPACE STUDIES OF SOLAR-SYSTEM OBJECTS: THE POST-ISO ERA

THÉRÈSE ENCRENAZ
DESPA, Observatoire de Paris, F-92195 Meudon, France

Abstract. The ISO mission is expected to allow significant progress in the study of Solar-System objects, especially concerning planetary and cometary atmospheres. Beyond ISO, future Solar-System studies using infrared space missions will require an extension of the spectral coverage toward longer wavelengths and increased spatial capabilities for imaging spectroscopy.

Key words: infrared astronomy, atmospheres, infrared spectroscopy.

1. Introduction

Solar-System objects, which have effective temperatures in the range of a few hundred K or less, are best suited for infrared studies, since most of their thermal energy is emitted in this wavelength range. In particular, infrared spectroscopy is ideally designed for the chemical study of the gaseous phase, through a large number of strong rotational and ro-vibrational transitions. For these reasons, many new results are expected to come out of the exploration of Solar-System objects with the ISO mission. A description of the capabilities of the ISO satellite, which will explore the infrared sky from Earth orbit in 1994–1995, can be found in Kessler (1985) and in Metcalfe and Kessler (1991).

In this paper, the major goals and limitations of ISO Solar-System observations will be briefly reviewed, and suggestions will be made for new satellites and instruments in the post-ISO era.

2. Terrestrial Planets

Because of the solar elongation angle constraint of ISO ($> 60°$), observations of Venus – and *a fortiori* Mercury – will not be feasible with ISO. The only observable terrestrial planet will thus be Mars.

In spite of an intense space exploration, the atmosphere of Mars is still far from being fully understood. Basic questions such as the stability of the atmosphere and the recycling of CO_2 (Atreya and Blamont, 1990), the nature of the surface, the possible presence of H_2O permafrost under the surface (Squyres, 1989) are still open. Because of the high inclination angle of its rotation axis, seasonal effects are very strong, involving the formation of polar caps and transport of H_2O and CO_2 from a pole to another. For all these reasons, and for its analogy with the planet

Space Science Reviews **61:** 13–17, 1992.

Earth, Mars, recently revisited by the Phobos 2 mission (see the special issue of *Plan. Space Science,* **39**, 1991) is going to be a major objective for planetary *in situ* space exploration.

ISO will allow us to search for minor species with high resolving power, using the Fabry-Pérot mode of the two spectrometers. These minor species (for instance chlorine compounds) are likely to play a key role in the CO_2 photodissociation and could be involved in the mechanisms which ensure its stability. Mars is bright enough for a complete spectrum to be recorded in the FP mode ($R = 10^4$) in a few hours integrating time. ISO limitations will be of two kinds: first, the cut-off at 180 μm will prevent the observations of a large number of species, which show rotational signatures in the submillimetre range (Encrenaz *et al.*, 1991); second, the small size of the ISO telescope does not allow the mapping of the martian disk below 10 μm with enough spatial resolution. SIRTF will not help for these studies, because its spectral resolving power (10^3) will be too small in the far-IR and submillimetre range, and the size of its telescope is not significantly larger than ISO for near-IR studies.

Future Earth-orbit IR observations of Mars should thus be extended in the two following directions: (1) extension of the spectral range toward 1 mm, with a resolving power of at least 10^4, and (2) use of a larger telescope for medium spectral resolution (10^3) mapping.

The first objective should be achieved with FIRST which is expected to provide a complete spectral survey in the 200 μm–1 mm range with the spectral resolving power of heterodyne spectroscopy, i.e. 10^6. The second objective could be met with an EDISON-type satellite. The 2.5m telescope would provide a spatial resolution of about 300 km, sufficient, for instance, to detect the CO spatial variations recently suggested by the ISM-PHOBOS experiment (Rosenqvist *et al.*, 1991). The best instrument to monitor simultaneously several minor species below 12 μm, would be a 2-D camera equipped with a grating, giving a resolving power of about 10^3. With this instrument, the simultaneous observation of CO_2, CO, H_2O, O_3... would be possible.

As a conclusion, one should mention that, for an IR satellite of the next generation, it would be important to reduce the solar elongation angle constraint to less than 40°, to allow the observation of Venus.

3. Giant Planets

Thanks to an intense exploration campaign over the past 15 years, both from the ground and from the Pioneer and Voyager spacecrafts, we have a good knowledge of the atmospheres of the giant planets (see the special issues of *Science*, **204**, 945-1008 (1979), **212**, 159-243 (1981), **215**, 499-594 (1982), **233**, 39-109 (1986) and **246**, 1417-1501 (1989). This exploration, in turn, has raised important new questions. Among them, 3 appear of particular importance (Encrenaz, 1990a): (1) the study of the early stages of Solar System evolution, through the measurement of elemental and isotopic ratios; (2) the study of the general circulation of the giant planets, through the monitoring of disequilibrium species, tracers of vertical motions (PH_3, AsH_3, CO, GeH_4...); the photochemistry and the study of aurorae, where complex molecules might be found. These questions will be among those to be addressed by the future space missions devoted to the giant planets, Galileo toward Jupiter and Cassini toward Saturn.

ISO will be a powerful tool for examining minor species, especially with the long-wavelength spectrometer in the FP mode. In particular, the high spectral resolution will allow a retrieval of the vertical distribution of the observed species, if present above the tropopause. In particular, a good measurement of D/H is expected from the study of the rotational HD lines (Bezard *et al.*, 1986). However, limitations are the same as in the case of Mars. As discussed by Bezard *et al.* (1986), several interesting species have signatures beyond 180 μm (HCP, H_2Se...), so that the spectral range to be covered should be enlarged. In addition, as for Mars, the spatial resolution of ISO for Jupiter and Saturn is going to be very low.

As for Mars, the 2 directions to work in the future are (1) an extension of the spectral coverage toward 1 mm, which should be achieved with FIRST, and (2) an EDISON-type satellite with a 2.5 m class telescope, giving a spatial resolution of 40 pixels over the Jovian meridian. The instruments to be used on this satellite would be the 2-D camera mentioned above, equipped with a grating ($R = 10^3$) and a high-resolution spectrometer below 14 μm (not available on ISO).

4. Satellites with a Gaseous Atmosphere

The best candidate is Titan, Saturn's satellite, which will be a major objective of ISO Solar-System observations. Titan has been explored with Voyager 1 (see the special issue of *Science*, **212**, 159-243), which has revealed its nitrogen atmosphere and has detected a large number of complex molecules. Among the questions to be addressed by the Cassini mission, are the nature of the surface (solid or liquid), the existence and the nature of the ocean, the nature of the interior, and the possible development of prebiotic chemistry (Coustenis, 1991).

ISO will search for minor constituents, will try to measure isotopic ratios (in particular D/H from CH_3D) and will determine vertical atmospheric profiles of minor species. The high spectral resolution is here especially useful, since the lines formed in emission in the stratosphere are much narrower than in the case of the giant planets (Coustenis *et al.*, 1991). ISO limitations will be, as for the previous cases, the cut-off at 180 μm but also the absence of FP mode below 14 μm (which would be important for the study of several hydrocarbons showing ro-vibrational bands in the $10 - 15$ μm range), and the limited sensitivity at 10 μm.

Future studies will include a spectral coverage in the submillimetre range with FIRST, and an EDISON-type satellite with both a high-resolution spectrometer ($R = 10^4$) and a medium resolution one ($R = 10^3$) below 14 μm.

Besides Titan, other satellites have been found to have a tenuous atmosphere:

Triton and Io. Triton is too weak to be observable with ISO in less than a few hours, but would be a very attractive target for a satellite like EDISON, with a grating/array spectrometer with moderate spectral resolution (R = 100) in the reflected sunlight component (2 – 10 μm). Io will be observed with ISO in the near-infrared range, but far-IR observations of its atmosphere will be also attempted with FIRST, after the first detection of SO_2 on Io from heterodyne spectroscopy (Lellouch *et al.*, 1990).

5. Comets

Because they are cold and small objects, comets are believed to be unaltered remnants of the early stages of the Solar System history. The major questions of cometary physics concern the nature of the nucleus (which can be probed through the study of the outgassed parent molecules), the thermodynamics of the coma (to be explored from the rotational structure of molecular emissions), and the link between cometary and interstellar material (in particular through the hydrocarbon emission at 3 μm; see Encrenaz, 1990b).

ISO studies of comets will include the search for parent molecules, through their emissions in the near IR (2 – 5 μm) and the far-IR range (beyond 100 μm), and the study of cometary dust. ISO limitations will come from the cut-off at 180 μm (since many possible candidates have strong rotational signatures in the submillimetre range, as discussed in Crovisier, 1987) and the sensitivity limit due to the small size of the telescope. Another ISO limitation, specific to comets, is the strong solar elongation angle constraint: it will be very difficult to sequence repeated observations of a single comet at various heliocentric distances.

Future IR observations of comets from Earth orbit will include observations at high resolving power in the submillimetre range with FIRST, and observations in the near IR with an EDISON-type satellite. In the latter case, the size of the telescope will allow us to observe a number of periodic comets larger than in the case of ISO (about 5) by at least an order of magnitude. If the solar elongation angle constraint can be reduced, repeated observations will be possible, taking also into account the extended lifetime of the mission.

6. Conclusion

As mentioned above, the ISO mission is expected to allow a very significant progress in our understanding of planetary and cometary atmospheres. Future space missions should concentrate along two main directions: [1] an extension of the spectral coverage toward longer wavelengths and [2] an increase in spatial resolution and in sensitivity, with the use of a larger telescope, allowing imaging spectroscopy with a resolving power of at least 10^3. In both cases, it will be essential to keep the solar elongation angle constraint as low as possible, in order to allow observations of terrestrial planets and comets.

References

Atreya, S.T. and Blamont, J.E.: 1990, *Geophys. Res. Letters*, **17**, 287.
Bezard, B., Gautier, D. and Marten, A.: 1986, *Astron. Astrophys.*, **161**, 387.
Coustenis, A.: 1991, *Vistas in Astronomy*, **34**, 11.
Coustenis, A., Encrenaz, T., Bezard, B., Kunde, V., Bjoraker, G., Graner, G., Dang-Nhu, M. and Arie, E.: 1991, submitted to *Icarus*.
Crovisier, J.: 1987, *Astron. Astrophys. Suppl. Series*, **68**, 223.
Encrenaz, T.: 1990a, *Rep. Prog. Phys.*, **53**, 793.
Encrenaz, T.: 1990b, *ESA*, **SP-315**, 127.
Encrenaz, T., Gulkis, S. and Lellouch, E.: 1991, in *"Coherent Detection Techniques at Millimetre Wavelengths and their Applications"*, Nova Science, publisher.
Kessler, M.F.: 1985, Proceedings of the Conference *"Instrumentation for Optical Remote Sensing from Space"*, Cannes, Nov. 1985, Proc. SPIE, **589**.
Lellouch, E., Belton, M., de Pater, I., Gulkis, S. and Encrenaz, T.: 1990, *Nature*, **346**, 639.
Metcalfe, L. and Kessler, M.F.: 1991, *"Scientific Capabilities of the ISO Payload"*, ESA-ESTEC Publication ISO-SSD-8805, August 1991.
Rosenqvist, J., Drossart, P., Combes, M., Encrenaz, T., Lellouch, E., Bibring, J.P., Erard, S., Langevin, Y. and Chassefiere, E.: 1991, submitted to *Icarus*.
Squyres, S.W.: 1989, *Icarus*, **79**, 229.

THE INTERSTELLAR MEDIUM AND STAR FORMATION

PETER W.J.L. BRAND
University of Edinburgh, Department of Astronomy, Blackford Hill, Edinburgh EH9 3HJ, U.K.

1. Star Formation

At the time of writing, the 'Holy Grail of Infrared Astronomy' (Wynn-Williams, 1982), i.e. the unequivocal identification of a protostar, has yet to be achieved. However, even this is the least of the problems of those who seek a successful theory of star formation. The protostar phase is important, but is only a small part of the full cyclic and multiply-fed-back system which regulates the conversion of gas to stars. This has been summarised in the important paper by McKee (1989).

The cycle can be traced from molecular cloud aggregation, energised by supernova blasts, of previously fragmented clouds and intercloud material. When large enough – giant – molecular clouds have accumulated, supported by magnetic fields and internal energy sources, star formation can begin. This proceeds by ambipolar diffusion in the neutral cloud cores, on a timescale depending only on the degree of ionization x of the material ($t_{\rm AD} = 3 \times 10^{14} x$ y) (Shu *et al.*, 1987). The degree of ionization is low enough that significant collapse occurs only in the densest cores (a few percent of the mass of the cloud) where ionization is caused solely by cosmic rays. The stellar UV flux, which would otherwise ionize metals, is completely excluded.

Since cloud contraction increases optical depth, contraction increases the fraction of the cloud susceptible to collapse to stars. It appears (Lada, 1985) that all forming stars produce a molecular outflow driven presumably by violent winds from either the stellar surface, or a surrounding disk, or both. This outflow is observed as (a) rapidly moving gas, and (b) shocks, of which more in section 2. The outflow dumps energy and therefore acts against cloud contraction. The opposition of increasing opacity, and the consequently increasing resistance to contraction, creates a balance between collapse and expansion that determines the star formation rate.

The eventual creation of a massive (say $10 M_\odot$) star causes the destruction of the cloud by the formation of an HII region which rapidly erodes the remaining cloud (Whitworth, 1979). The mechanical power output of supernova remnants and the interstellar UV flux are determined by the initial spectrum of stellar masses.

Such a scheme (McKee *ibid.*) is able to predict with satisfactory accuracy the overall rate of star formation in our galaxy; is the scheme correct so far? Will the predictions of other parameters prove equally successful? Can we extrapolate to

Space Science Reviews **61:** 19–23, 1992.

significantly different situations (dwarf galaxies, ellipticals, the early Universe)? Much remains to be done.

Two critical phenomena in the scheme are observable. These are: the injection of kinetic energy into the molecular cloud by the supersonic outflows; and the irradiation of molecular clouds by the UV radiation from newly formed massive stars. These phenomena occur with very high energy densities, and therefore emit at relatively short wavelengths, where the current IR technology allows detailed investigation.

I shall describe our investigations of these two processes in sections 2 and 3. We have used measurements of the emission lines of excited molecular hydrogen to investigate each of them.

2. Shocks and Molecular Outflows

The rotational/vibrational lines of excited molecular hydrogen (the dominant species) fall conveniently in the near IR spectrum (1-5 μm). This region is ideal because dust extinction is less by an order of magnitude than that in the visible spectrum (visual optical depths of approximately 8 or more are expected in star forming regions, McKee *ibid.*). Furthermore, this region is now well catered for by the new spectrometers, notably these at the UK Infrared Telescope.

These H_2 lines are excited by the high (greater than 2000K) temperatures behind shocks in the molecular gas. Proper interpretation of the line ratios and intensities provides important quantitative information about the outflow (wind pressure or momentum flux) and about local properties (density, chemical state).

As an example, I shall describe our work on the outflow in Orion, driven by the prestellar source IRc2.

The normalised column densities of the upper level of the H_2 transition, derived from the line intensities, form the basis of the analysis. The densities are 'normalised' by division by the statistical weights. If the levels were thermally populated at some temperature T then the slope of the logarithm of normalised column density versus energy would be $-1/kT$.

The results from Orion show that the normalised column density versus energy indeed forms a smooth curve, but that the slope changes with energy. The higher the energy of the level, the higher the apparent temperature derived from the slope.

A simple explanation of this phenomemon is that we are looking at the cooling zone behind a shock (Brand *et al.*, 1988). The predicted values from such a plane parallel shock can be made to fit the data precisely.

Two virtually model-independent conclusions can be drawn. These follow from the smoothness of the column density/energy curve.

The first is that the lines must be collisionally populated, implying high densities in the shocked gas. The second arises from the nature of extinction corrections. Lines from transitions from similar (or identical) upper energy levels can occur at quite different wavelengths, and therefore be subject to quite different extinctions.

The best estimate of extinction is therefore the one that reduces the scatter of the corrected column densities, independent of the shape of the underlying curve, and is therefore rather well-determined because of the large number of lines involved.

A more model-dependent conclusion (it depends on the assumption that a cooling zone is being observed) is that the shape of the column density curve depends on the cooling function, and therefore on the major coolants and on the pressure driving the shock.

If, on the other hand, a magnetically-moderated shock (C-shock, Draine 1980) – with heating and cooling occurring simultaneously – is being observed, then the shape of the curve (column density versus level energy) is determined to a large extent by the bow structure of a non-plane shock. Planar C-shocks are ruled out. However we (Smith *et al.*, 1991) have shown that the observations at the brightest part of the Orion outflow can be replicated by a bow-shaped C-shock, with rather particular constraints.

These rather complex results are further complicated by the data on the velocity profiles of the H_2 lines. Their great widths (Nadeau and Geballe, 1979; Brand *et al.*, 1989) imply relative velocities of up to 140 km s^{-1}, while it is well-known that non-magnetic shocks in H_2 dissociate the gas completely if the shock speed is greater than 25 km s^{-1} (Kwan, 1977), while C-shocks do so for velocities greater than 40-50 km s^{-1} (Draine and Roberge, 1982). Smith *et al.* (*ibid.*) suggest very high magnetic field C-shock structures may explain the data, but there are other complications.

These studies are outlined here to emphasise the fact that new spectroscopic instrumentation in the infrared can so constrain the models of outflows as to yield – eventually – important parameters for the outflows taken to power star-forming cores.

3. Photodissociation Regions

Since the recent resurgence of interest in the UV-irradiated boundaries of molecular clouds termed photodissociation regions (Tielens and Hollenbach, 1985), observations of these regions have been made at a wide variety of wavelengths (e.g. Genzel *et al.*, 1989). The molecular hydrogen emission in the infrared is stimulated by UV excitation of an electronic level which cascades through the vibrational/rotational levels of the ground electronic state. Approximately ten percent of such transitions result in a dissociation of the molecule, whose destruction is balanced by hydrogen molecule formation on the surfaces of dust grains. This gives a statistical equilibrium, assumed in all such models so far.

The photodissociation region exists as a boundary layer between HII regions and neutral molecular clouds (Genzel, *ibid.*), or at the surfaces of reflection nebulae (Gatley *et al.*, 1987). Extensive theoretical studies (Burton, Hollenbach and Tielens, 1990; Strenberg and Dalgarno, 1989) have demonstrated that these regions can be quite hot, well over 1000 K.

These energetic sources are powerful emitters of radiation, and at wavelengths longer than, say, ten micrometres their emission is little impeded by dust absorption.

It is in the nature of star formation that these interfaces – requiring as they do the conjunction of dense clouds and massive stars – will tend to be deeply embedded in star forming regions, and therefore optically undetectable. Their powerful infrared emission, which we can relate to well-found physics in the photodissociation region, make them ideal devices to probe star forming regions (perhaps most particularly starburst regions in other galaxies).

One necessary qualification is our physical understanding of these regions. They have been well studied theoretically, but observational work to check the theory is still in its infancy.

We have observed photodissociation regions in Orion, NGC2023 and M17. Our aim has been to understand the hydrogen spectrum which (for these relatively nearby regions) enables us to determine the geometry of the source, and through line ratios, the density. These and related studies will ensure that we understand the physics, and can use these sources as probes of star forming regions.

A particular quest is to check the assumption that these configurations are static, with formation and destruction rates of molecules of hydrogen equal. There is an indication in our data that the intensity in NGC2023 is higher than that which can be derived from the models. If this is the case, it may imply that the region is venting dissociated gas (similar to a 'blister' HII region), leaving fresh material to be irradiated by UV starlight.

If this turns out to be true, then photodissociations take their place alongside HII regions as destroyers of molecular clouds.

4. The Future

These remarks on shocks and photodissociation regions are made to show that the new observing technologies in the infrared enable fundamental observational tests of theories, and can provide the parameters these theories urgently require.

Progress on understanding the 'hot' molecular gas has been rapid, and now vies with our comprehension of cold gas in the interstellar medium, the province of radio astronomy.

On reflection it is clear from a star formation scheme like that described by McKee (*ibid.*) that many of the important energy transfers take place at *intermediate* temperatures, in the range of hundreds of degrees Kelvin. These phenomena include heating by ambipolar diffusion, by turbulent diffusion, by transfer of angular momentum, and by dissipation of Alfven waves. Much more can be learned, also, about the 'hot' gas regions from their warm surroundings.

Thes phenomena are largely inaccessible to us, since the characteristic emission wavelengths lie at tens to hundreds of micrometres where the atmosphere is opaque. But a bright new future – and possibly at last a verification of a star formation *theory* – is heralded in by the young and vigorous science of infrared satellite astronomy.

Acknowledgements

This work is based on a collaboration with many co-workers, notably T.R. Geballe, M.G. Burton, A. Moorhouse, M.D. Smith, A.S. Webster, M. Toner, A. Chrysostomou. I most especially thank the UK Infrared Telescope staff, scientific, technical and administrative, for making it easy to obtain wonderful data, and I thank the Science and Engineering Research Council for financial support.

References

Brand, P., Moorhouse, A., Burton, M., Geballe, T., Bird, M. and Wade, R.: 1988, *Astrophys. J. (Lett.)*, **334**, L103.

Brand, P., Toner, M., Geballe, T. and Webster, A.: 1989, *Mon. Not. R. astr. Soc.*, **237**, 1009.

Burton, M., Hollenbach, D. and Tielens, A.: 1990, *Astrophys. J.*, **365**, 620.

Draine, B.: 1980, *Astrophys. J.*, **241**, 1021.

Draine, B. and Roberge, W.: 1982, *Astrophys. J. (Lett.)*, **259**, L91.

Gatley, I., Hasegawa, T., Suzuki, H., Garden, R., Brand, P., Lightfoot, J. and Glencross, W.: 1987, *Astrophys. J. (Lett.)*, **318**, L75.

Genzel, R., Harris, A. and Stutzki, J.: 1989, *22nd ESLAB Symposium*, ed. Kaldeich.

Kwan, J.: 1977, *Astrophys. J.*, **216**, 713.

Lada, C.: 1985, *Ann. Revs. Ast. and Astrophys.*, **23**, 267.

Nadeau, D. and Geballe, T.: 1979, *Astrophys. J. (Lett.)*, **230**, L169.

McKee, C.F.: 1989, *Astrophys. J.*, **345**, 782.

Shu, F., Adams, F. and Lizano, S.: 1987, *Ann. Revs. Ast. and Astrophys.*, **25**, 23.

Smith, M.D., Brand, P. and Moorhouse, A.: 1991, *Mon. Not. R. astr. Soc.*, **248**, 730.

Sternberg, A. and Dalgarno, A.: 1989, *Astrophys. J.*, **338**, 197.

Tielens, A. and Hollenbach, D.: 1985, *Astrophys. J.*, **291**, 722.

Whitworth, A.: 1979, *Mon. Not. R. astr. Soc.*, **186**, 59.

Wynn-Williams, C.G.: 1982, *Ann. Revs. Ast. and Astrophys.*, **20**, 587.

STELLAR INFRARED ASTRONOMY WITH EDISON

L. B. F. M. WATERS
SRON Laboratory for Space Research,
P.O. Box 800, 9700 AV Groningen, The Netherlands

Abstract. In this review the IR emission from circumstellar material is discussed, both of ionized gas and dust grains, and the astrophysical information that can be extracted from such observations. Some emphasis is placed on the possibilities of stellar IR astronomy using a large space-borne telescope, especially with respect to the much better spatial and spectral resolution of such a telescope compared to the current generation of ground-based and space IR telescopes.

Key words: Infrared radiation – stars: circumstellar material.

1. Introduction

The recent improvement in IR observing facilities, both ground-based and space-borne, has resulted in exciting new results in all fields of astronomy, ranging from solar system studies to cosmology. In particular the IRAS all-sky far-IR survey at 12, 25, 60 and 100 μm has provided a unique database for the study of the IR properties of various types of objects. In the field of stellar IR astronomy, which is the topic of this review, IRAS and follow-up observations (based on IRAS data) have dramatically increased our knowledge of circumstellar material, mass loss rates and their impact on stellar evolution. New questions that were raised by the IRAS observations often require a (much) higher spatial and/or spectral resolution than could be achieved with IRAS. Such questions will undoubtedly be addressed by the upcoming generation of IR space observatories: the Infrared Space Observatory ISO and the Space Infrared Telescope Facility SIRTF. While many problems in astrophysics can be tackled with these observatories, they do have their limitations, especially in terms of spatial resolution (due to the relatively small mirrors for both ISO and SIRTF of 0.6 and 0.9 m respectively) and spectral resolution in combination with sensitivity. Both ISO and SIRTF will have a spectral resolution of the order of 10^3, while ISO offers the possibility to observe with the Fabry-Pérot interferometers at a resolution of 10^4 to 3 10^4.

In this review I try to look beyond ISO and SIRTF (!) to discuss some problems in astrophysics that may be addressed with the next generation of IR space observatories, such as e.g. EDISON (described elsewhere in this volume). I will concentrate on those aspects that *cannot or can only partly* be done with ISO and SIRTF: observations that require high spatial and/or spectral resolution. Inevitably I will be biased by my own interests, and this review obviously cannot cover all areas of astrophysics. In particular I will not mention star formation and brown dwarfs. In Section 2 I briefly discuss IR emission from circumstellar material, and

Space Science Reviews **61:** 25–43, 1992.

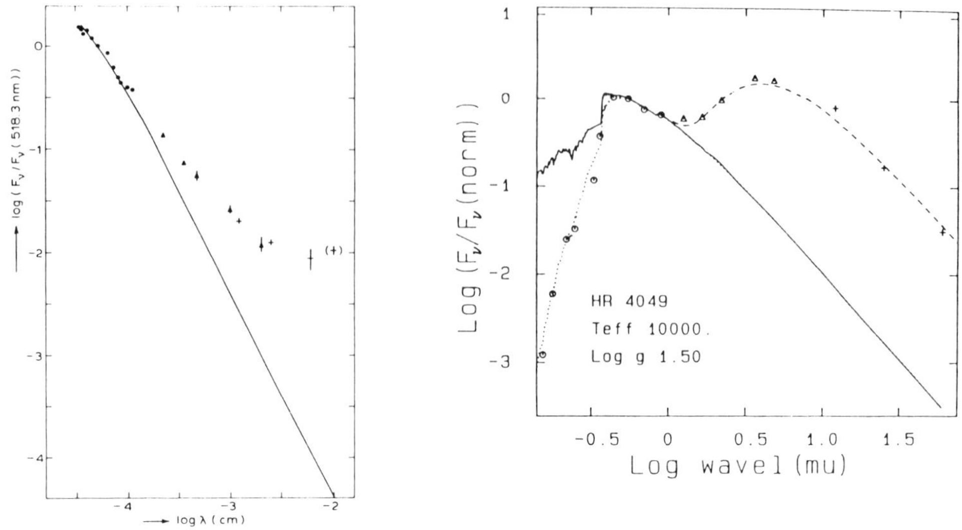

Fig. 1. Typical example of a star with IR excess due to free-free emission from a dense stellar wind (the LBV P Cygni; Waters and Wesselius, 1986), and of a hot dust shell around a star (the post-AGB star HR 4049, Lamers *et al.*, 1986).

subsequent Sections deal with some areas in astrophysics to which a large IR space mission could contribute significantly. Section 6 summarizes and lists some recommendations for the type of focal plane instruments that are considered neccessary or desirable.

2. IR Emission from Circumstellar Material

Stars in various evolutionary stages show evidence for circumstellar material, either gas or dust. If the gas is ionized, it will emit free-free and free-bound radiation, which can be observed as an excess of radiation with respect to the stellar photospheric radiation. In the case of dust, the grains absorb stellar optical and UV radiation and re-emit this in the IR as thermal radiation at their equilibrium temperature. Since in general hot stars have ionized circumstellar material, they usually show free-free emission and recombination line radiation. In cool stars the material can condense into grains that emit thermal radiation, and also molecules can be observed in the photosphere and wind (e.g. CO, C_2H_2, SiO, HCN). In Figure 1 a typical example of a star with free-free emission and of thermal emission from grains is shown. The shape of the continuum energy distribution of the free-free and the dust excess is very different, which makes it easy to distinguish between them. Below we discuss some of the diagnostic tools that are available in the IR to study circumstellar material.

2.1. Free-Free Emission

When a star is surrounded by ionized gas, this will emit free-bound and free-free radiation, resulting in an excess of (far-IR) radiation. The absorption coefficient for bound-free and free-free emission can be written as:

$$\kappa_\nu = 3.69\ 10^8 (h/k) Z^2 (g + b) T^{-3/2} \nu^{-2} \gamma n_i^2 \tag{1}$$

where Z^2 is the mean value of the squared atomic charge, T is the electron temperature in K, n_i the ion number density, ν the frequency in Hz, γ the ratio of the electron and ion number density, and g and b are the gaunt factors for free-free and free-bound radiation respectively. If for simplicity we assume a power-law density distribution of the shape

$$\rho(r) = \rho_0 (r/R_*)^{-n} \tag{2}$$

a simple relation is found between the slope of the IR continuum and the density gradient of the gas (or the parameter n):

$$F_\nu \propto \nu^{(4n-6)/(2n-1)} \qquad (n \neq 0.5) \tag{3}$$

where we have neglected the (weak) wavelength dependence of the gaunt factor. Eq. (3) only holds for cases with a significant IR excess, i.e. when part of the circumstellar material has optical depth > 1. For $n = 2$ we find the familiar result $F_\nu \propto \nu^{2/3}$, which indeed is observed in the radio spectrum of massive supergiants (e.g. Bieging *et al.*, 1989). So the slope of the spectrum gives the density gradient of the gas (and from that, using mass continuity, the velocity law), while the size of the IR excess is a measure of the densities.

2.2. Thermal Radiation from Circumstellar Dust

In cool stars with mass loss (e.g. Mira's, OH/IR stars), but also in some hotter stars with very dense circumstellar shells, dust can condense in the outflow. This dust will absorb stellar optical and UV flux and re-emit it in the IR at some temperature. Here for simplicity we assume that the dust is optically thin. Assume furthermore that the grains are distributed spherically symmetric around the star with a number density distribution $n_d(r) = n_0 (r/R_*)^{-m}$ and a dust emissivity $Q_\nu = Q_0 (\nu/\nu_0)^p$. It can be shown that for optically thin dust in thermal equilibrium with the stellar radiation field:

$$T(r) = T_0 (r/R_*)^{-2/(4+p)} \tag{4}$$

with T_0 some normalisation temperature. When integrating the contribution from all grains between the inner and outer radius of the dust shell (r_0 and r_1 respectively), a simple relation can be found for the slope of the spectrum if $r_1 \gg r_0$:

$$F_\nu \propto \nu^{3+p-0.5(3-m)(4+p)} \tag{5}$$

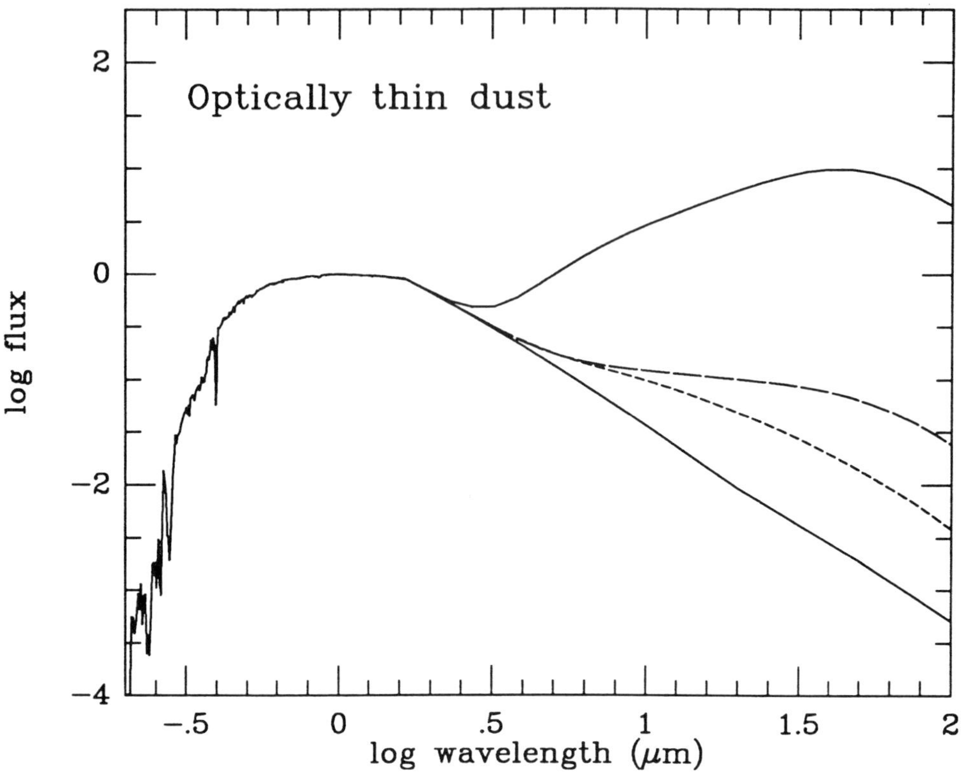

Fig. 2. Theoretical dust energy distribution for optically thin dust in thermal equilibrium with the stellar photosphere (Kurucz model, indicated as the solid line), and three different density gradients, namely $m = 1$ (solid line), 1.5 (long dash) and 2.

TABLE I
Some dust features observed in circumstellar shells

Wavelength (μm)	ident.
9.7, 18	Silicates
11.2	SiC
3.08, 6.0	H_2O ice
3.3, 7.7, 8.6, 11.3	UIR bands

In Figure 2 we show some theoretical dust energy distributions with different choices of the density gradient of the dust grains (parameter m), for a dust cloud illuminated by a central star with temperature T_{eff} = 5500 K. As can be seen, the slope of the spectrum is very sensitive to the density gradient.

Eq. (5) shows that several combinations of the parameters p and m can in principle result in the same slope of the dust continuum. Due to our poor knowledge of the wavelength dependence of the grain emissivity (parameter p), it is not possible to derive a unique value for the density gradient m from the slope of the spectrum alone. This ambiguity can be solved by spatially resolving the dust emission at different wavelengths, which allows an independent determination of p and m. We will come back to this point in a forthcoming Section.

Apart from continuum radiation, the dust also shows solid state features, that give information on the chemical composition of the dust. Well-known examples are the 9.7 and 18 μm silicate features, indicative of oxygen-rich dust, and the 11.2 μm SiC feature, which is attributed to carbon-rich dust. In Table I we list some of the dust features with their identification (taken from Roche, 1988).

2.3. IR Recombination Lines and Fine-structure Lines

Apart from emitting IR continuum radiation, the circumstellar gas also emits line radiation. If the gas is sufficiently ionized, the recombination lines of H and He can be observed, as well as forbidden line transitions from many atoms and ions (e.g. C, N, O, Si, Fe, Ne, S, etc.). Both the recombination lines and the fine structure lines are important diagnostics for the physical conditions in the circumstellar environment.

The recombination lines of H and He give information on the densities and velocities in their region of formation. If the spectral resolution is insufficient to resolve the lines, the equivalent width, and line strength ratios give information on the densities and density gradients of the ionized gas.

Following Castor (1970) and Simon *et al.* (1983) we can derive a general expression for the line profile of hydrogen recombination lines in a spherically symmetric wind with a power-law velocity law of the shape $v(r) = v_0(r/R_*)^{n-2}$. We assume for simplicity that we can neglect the underlying photospheric line profile (a reasonable assumption for hot stars in the mid- and far-IR), and we also assume that the source function in the wind is not too different from that of the photosphere (which is the Planck function). These assumptions allow us to neglect the column of gas projected against the disc of the star. We find:

$$F_L(\nu) = R_*^2/D^2 \int_{x_{min}}^{\infty} S_\nu(1 - e^{-\tau_\nu(x)})(1 + \beta^2(n-3)x^{4-2n})2\pi\, x\, dx \qquad (6)$$

where $\tau_\nu(x)$ is the optical depth along a line of sight, D is the distance in cm, S_ν the line source function, ν_0 the frequency at line centre, and β and x_{min} are given by:

$$\beta = \frac{c(\nu - \nu_0)}{\nu_0 v_0} \qquad (7)$$

$$x_{min} = 1 \quad for\ \beta < 1 \qquad (8)$$

$$x_{min} = \beta^{1/(n-2)} \quad for\ \beta > 1\ and\ n \neq 2 \qquad (9)$$

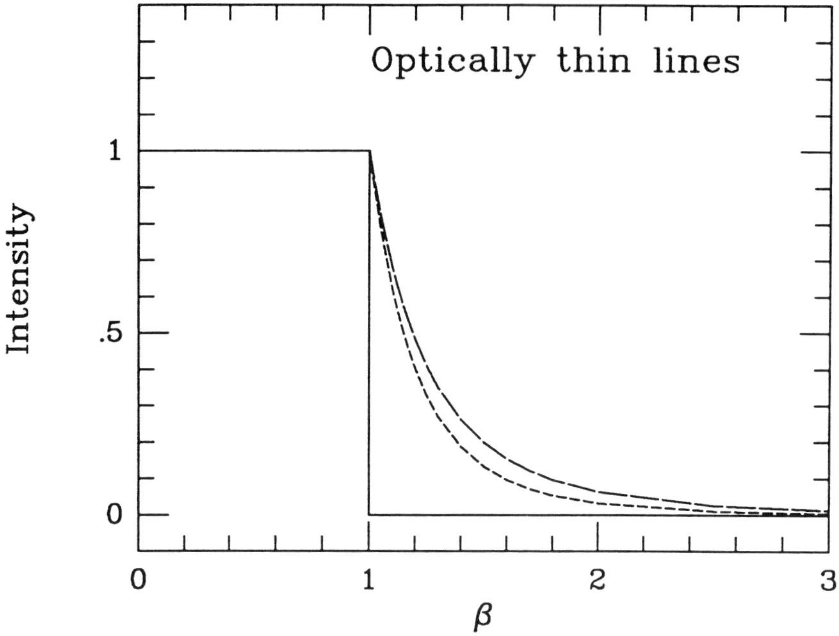

Fig. 3. Theoretical line profiles of HI recombination lines in a spherically symmetric, optically thin wind with a power-law velocity law $v(r) \propto r^{n-2}$, for three values of n, namely $n = 2$ (solid line), 2.5 (short dash), and 3 (long dash). Notice the effect of n on the line wing, for $\beta > 1$.

The optical depth $\tau_\nu(x)$ is given by:

$$\tau_\nu(x) = \frac{Kx^{-3n+3}}{1+\beta^2(n-3)x^{4-2n}} \tag{10}$$

and the constant K:

$$K = \left(\frac{\pi e^2}{m_e c}\right)\left(\frac{h^2}{2\pi mkT}\right)^{3/2}\left(\frac{\dot{M}}{4\pi\mu m_H v_0 R_*^2}\right)^2 \times (b_l - b_u e^{-h\nu/kT})e^{-\chi_L/kT}\left(\frac{cR_*}{\nu_0 v_0}\right) gf \tag{11}$$

where $\dot{M}$ is the mass loss rate, gf the oscillator strength, χ_L is the energy of the lower level, μ the mean atomic mass in units of the proton mass, and b_l and b_u are the departure coefficients of the lower and upper levels respectively. Under the assumption of LTE, the source function S_ν is the Planck function, and the departure coefficients are unity.

In the case of optically thin lines, Eq. (6) reduces to:

$$F_L(\nu) = 2\pi SKR_*^2(3n-5)^{-1}D^{-2} \qquad (\beta < 1) \tag{12}$$

$$F_L(\nu) = 2\pi SKR_*^2(3n-5)^{-1}D^{-2}\beta^{\frac{-3n+5}{n-2}} \qquad (\beta > 1; n > 2) \tag{13}$$

Eqs. (12) and (13) show that for optically thin lines the line shape in the wings depends on the velocity gradient (parameter n). In Fig. 3 we show some line profiles calculated on the basis of Eqs. (12) and (13) for several values of the velocity gradient. The case $n = 2$ yields the familiar rectangular line profile (Sobolev, 1958). This simple example shows that the shape of the IR recombination lines give valuable information on the velocities (total line width) and the velocity gradient (shape of the line) in the wind. Furthermore, the total line flux in the optically thin case is proportional to the ratio $(\dot{M}/v_0)^2$.

The IR fine-structure lines can be used as accurate indicators of the abundances in circumstellar material, and also of the wind outflow velocities. The IR lines are easier to interpret than the optical fine-structure lines, because the IR lines are not very sensitive to the temperature, but only to the density of the gas, while the optical lines are sensitive to both density and temperature. Furthermore, for some elements as Fe, Mg and Si there are no suitable optical forbidden lines that can be used for abundance analyses. An excellent example of the use of fine-structure lines can be found in Barlow et al. (1988), who observed the mid-IR spectrum of the Wolf-Rayet star γ Vel, and used the [NeII] line at λ 12.81 μm to obtain the wind velocities and total Ne abundance in this evolved star.

3. Proto-planetary Discs

Perhaps the most exciting discovery of IRAS in stars was the detection of cool circumstellar material in Vega (Aumann *et al.*, 1984) and several other, apparently normal main sequence stars (Aumann, 1985; Gillett, 1986; Waters *et al.*, 1987; Walker and Wolstencroft, 1988). In Figure 4 we show the optical to far-IR energy distributions for the four stars discussed by Gillett (1986). The solid lines are Kurucz model fits to the optical spectrum (except for ϵ Eri where we used a model kindly provided to us by Dr. Gustafsson). The IR excess is apparent for wavelengths longward of 10 to 20 μm. These observations are interpreted in terms of a disc of solid state material with particle sizes significantly larger than the dust found in the interstellar medium. This disc is thought to be the remnant of the parent molecular cloud from which the star formed, and which could presently be the site of planet formation. Slow scans performed with IRAS on Vega showed that the source is extended in the in-scan direction with a size of about 20 arcsec (Gillett, 1986). A recent re-evaluation of the data indicates a larger size of about 32 arcsec (van der Bliek, private communication). This latter value is in better agreement with the size of the disc of 40 arcsec found by Harvey *et al.* (1984) with the KAO.

The star β Pictoris was studied in great detail following the IRAS discovery, because it also shows evidence for the presence of a gaseous disc, most likely situated inside the dust disc seen by IRAS. High resolution IUE observations indicated that material is falling into the star with velocities up to 400 km/s (Lagrange-Henri *et al.*, 1989), while recently also the presence of a weak *outflow* was discovered (Bruhweiler *et al.*, 1991). It is clear that of the Vega-type systems known up to now

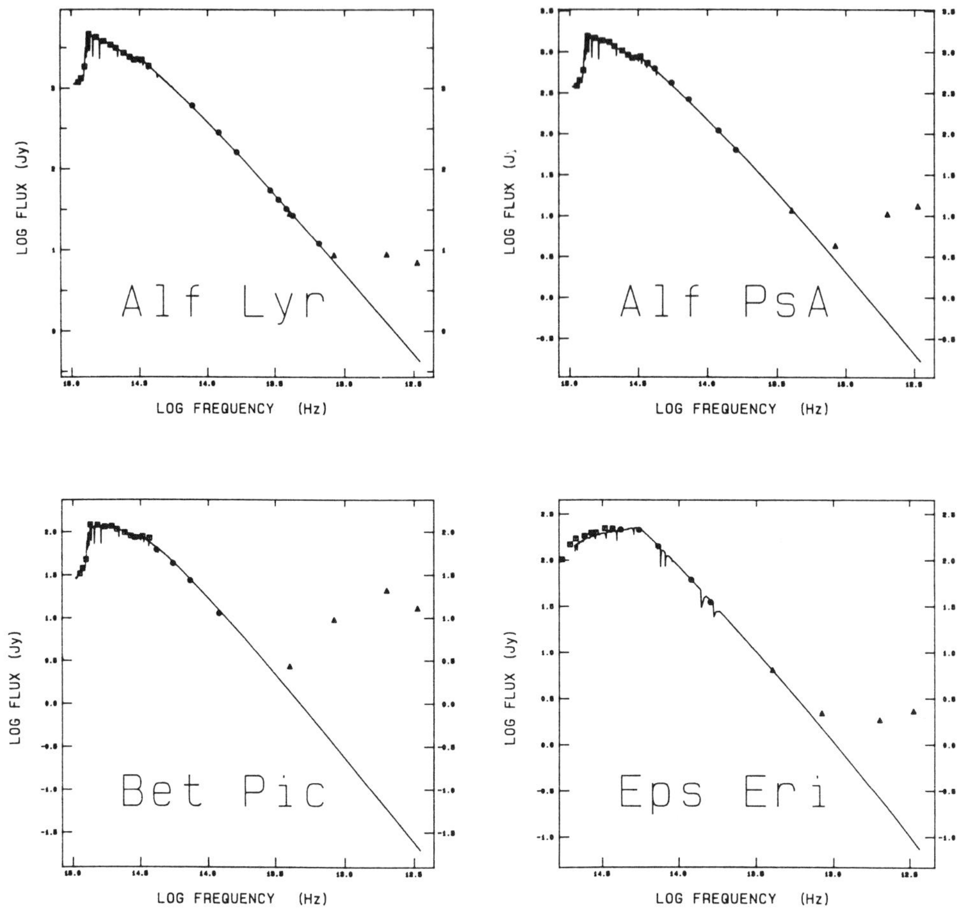

Fig. 4. Dereddened energy distribution of four well-studied Vega-type stars. The solid lines are the model atmospheres (Kurucz (1979) and Gustafsson (private communication)), the dots are the observed fluxes in the optical and IR. The models were fitted to the optical fluxes only. Notice the onset of the IR excess at wavelengths between 10 and 20 μm.

the β Pic system is by far the most active one, and together with 51 Oph has the warmest and most conspicuous dust shell. Apparently the β Pic disc is not as far evolved.

Important questions that arose after the first dust discs around main sequence stars were found, were on the frequency of occurrence of such discs, whether or not planet-size bodies are being formed, and how such discs evolve, perhaps into something comparable to the Oort cloud. Such questions could be answered by EDISON. Young open clusters can be used to investigate the frequency of dust shells around normal main sequence stars, by imaging these clusters in the K band and at 20 or 40 μm. Comparison of the K and far-IR brightness of the stars will immediately reveal those stars with IR excess. Comparison of a few clusters of different age will give insight into the evolution of such discs. For this a large IR telescope is needed with a spatial resolution of a few arcsec and sufficient sensitivity to detect G and K type dwarfs at 40 μm out to distances of the order of 1 kpc.

The spatial structure of the dust and gas around nearby Vega-type stars is another obvious area of investigation. By resolving the spatial structure of the dust discs, the ambiguity in the interpretation of the integrated dust energy distribution, as seen by telescopes with large apertures, can be resolved, i.e. the temperature structure and density distribution can be determined independently. This allows conclusions to be drawn on the grain size, and possibly changes in the grain size with distance from the star. These are important parameters for models that describe the evolution of such dusty discs. Spectral imaging of the gaseous part of the disc will reveal the spatial structure of the disc, and can check the interpretation of Lagrange-Henri *et al.* (1989) that the gaseous disc is largely due to evaporation of large comet-like bodies as they approach the central star.

4. Winds of Massive Stars

One may wonder why hot stars should be a topic for IR space astronomy, since hot stars emit most of their energy in the (far)-UV. However, much can be learned concerning the winds of hot stars by observing them in the IR. Both the continuum and the emission lines give valuable information on the density structure and the velocity field. This information cannot or can only partially be obtained by observations at other wavelengths (such as the UV). Furthermore, many massive stars are obscured by large amounts of interstellar dust, preventing detailed observations in the UV.

4.1. Mass Loss in Hot Stars

Mass loss is an essential ingredient of the evolution of massive stars, and we cannot even attempt to understand this evolution without a detailed knowledge of the mass loss rates and mechanisms in the various stages of evolution. Perhaps the best illustration of this effect is the observational HR diagram of the most massive stars in the Milky Way, SMC and LMC as presented by Humphreys and Davidson (1979), which shows a well-defined upper limit (the so-called Humphreys-Davidson limit), above which no stable star exists. Evolutionary calculations of stars with main sequence masses above about 50 $M_\odot$ without mass loss predicted that stars should exist in the 'forbidden' area of the HR diagram. The proper inclusion of mass loss in the evolutionary calculations by Maeder and Meynet (1987) provided much better agreement with the observed distribution of massive stars in the HR diagram.

An accurate knowledge of mass loss on the main sequence is important because this strongly influences the further evolution. If the main sequence mass loss is low, the star will lose a relatively small fraction of its envelope on the main sequence, and may not evolve into a Wolf-Rayet star. If the mass loss is very high, the star will evolve directly from the main sequence to the Wolf-Rayet phase, without becoming a red supergiant.

All evolutionary phases following the main sequence are dominated by large mass loss rates, of 10^{-6} $M_\odot$/yr and higher. Stars above about 50 $M_\odot$ will become very unstable as they evolve towards the red part of the HR diagram, and will eject large amounts of material during violent outbursts: the Luminous Blue Variables (LBV's). A well-known example of such a star is η Car. These outbursts result in the removal of most of the hydrogen-rich upper atmosphere, and nuclear processed material (CNO products) will be observable at the surface. Eventually the star will have a helium-rich atmosphere: it is a Wolf-Rayet star. Stars of lower main sequence mass will pass through the red supergiant phase where they shed most of the hydrogen rich envelope, and will also evolve to the Wolf-Rayet phase.

These events have great impact on the stellar energy distribution and the line spectrum, while in many cases the ejecta can also be observed directly. A good example is the ring nebula around AG Car (Nota and Paresce, 1989), which contains material ejected in a large outburst some 10^4 years ago. Other examples are ring nebulae around Pop. I Wolf-Rayet stars (Chu, 1991). The study of the chemical composition of the photosphere of the star, and of the ejecta gives strong constraints on the recent evolution of the star, and provides critical tests of evolutionary calculations that predict the change of surface chemical abundances with time.

The continuum energy distribution of hot stars in the IR is modified by the presence of the wind, since the ionized circumstellar gas will emit free-bound and free-free radiation (see Sect. 2). This emission originates from layers that are very close to the star, i.e. in which the rapid acceleration from very low velocities near the photosphere, to high terminal velocities (2000 km/s) occurs. By measuring the IR excess as a function of wavelength, information is obtained on the density distribution of the circumstellar gas. The study of this region gives important information on the mass loss mechanism in hot stars, most likely radiation pressure in UV resonance lines. Lamers and Waters (1984) applied this method to the B0 Ia supergiant ϵ Ori, and found that the velocity increases much more gradually than predicted by the radiation driven wind theory (e.g. Castor *et al.*, 1975). Bertout *et al.* (1985) investigated the IR excess in a large sample of massive stars, and concluded that the velocity gradient in the winds near the photosphere may become very shallow going from late O to late B spectral type, which again is not predicted by theory. It is then remarkable that the shape of the UV resonance lines observed in many stars is in good agreement with theory (e.g. Groenewegen and Lamers, 1989). However, the UV resonance lines do not give reliable information on the layers close to the photosphere, because in many cases the lines are saturated near zero velocity. Also the underlying photospheric line prevents detailed analysis of the velocity field close to the star. This wind region can thus only be studied using the IR excess method.

Recombination lines of hydrogen and helium, as well as fine structure lines of various atoms and ions, can be used to obtain information on the velocity field (Sect. 2). The shape of the lines is very sensitive not only to the details of the velocity law in the wind, but also to the densities. Lines of different strength are formed in different parts of the wind, thus providing probes of the wind structure in their formation region. The IR recombination lines are relatively easy to model, since the level populations of the high levels of the H atom are probably close to LTE. Furthermore, these lines are not very sensitive to the details of the temperature distribution in the wind. It is well known that the winds of hot stars contain some gas at very high temperatures (e.g. Cassinelli and Swank, 1983). This makes it extremely difficult to interpret the line strengths of UV resonance lines of trace elements as C IV and N V in terms of mass loss. The IR recombination lines are not affected by this, and should give reliable mass loss estimates.

4.2. Be Stars

A special subset of hot stars are the Be stars. These are main sequence stars of intermediate mass (2.5 to about 20 $M_\odot$) that show the Hα line in emission. They are on average rapid rotators, with $vsini$ typically 100-400 km/s, i.e. a significant fraction of the equatorial break-up velocity. The winds of Be stars are probably not spherically symmetric but *disc-like*, a fact which must be linked to the rapid rotation of the stars. In the equatorial regions, the wind is characterized by a high-density, low-velocity outflow, in which the motions are dominated by the rotation of the ejected gas around the star (still a few 100 km/s) rather than by the radial motion (10-100 km/s). At higher latitudes, a fast but tenuous wind emerges, which can be observed in the UV, and which reaches outflow velocities of 500-1500 km/s, i.e. well above the surface escape velocity. The physical mechanism for this highly variable mass loss is not understood. The theory of radiation driven winds including the effects of rotation does not predict winds in agreement with observations. The role of magnetic fields is not clear. A detailed investigation of the structure of the winds of these stars could clarify the situation.

In the IR the stars are characterized by a large excess due to free-free emission from the dense gas which also produces the Hα emission. This gas also emits IR recombination line radiation, which can be used to probe the velocity field of the equatorial material at various distances. This information cannot be obtained by observing the stars in e.g. the UV, because the lines that can readily be observed in that wavelength region are not formed in the high density regions close to the equator of the star. The Balmer lines could in priciple be used, but they are very difficult to model, and the star dominates the continuum emission in the optical.

In Figure 5 we show an example of the shape of IR recombination lines that might be expected from a typical Be star, based on a model developed by J.M. Marlborough. The model consists of a B0 star with a cone-shaped disc with opening angle 5 degrees seen edge-on. Inside the cone the gas has a density distribution $\rho(r) = \rho_0(r/R_*)^{-3}$, with $\rho_0 = 5\ 10^{-12}$ g/cm^3 (we ignore material outside the cone). This radial density distribution corresponds to an outflow velocity of the

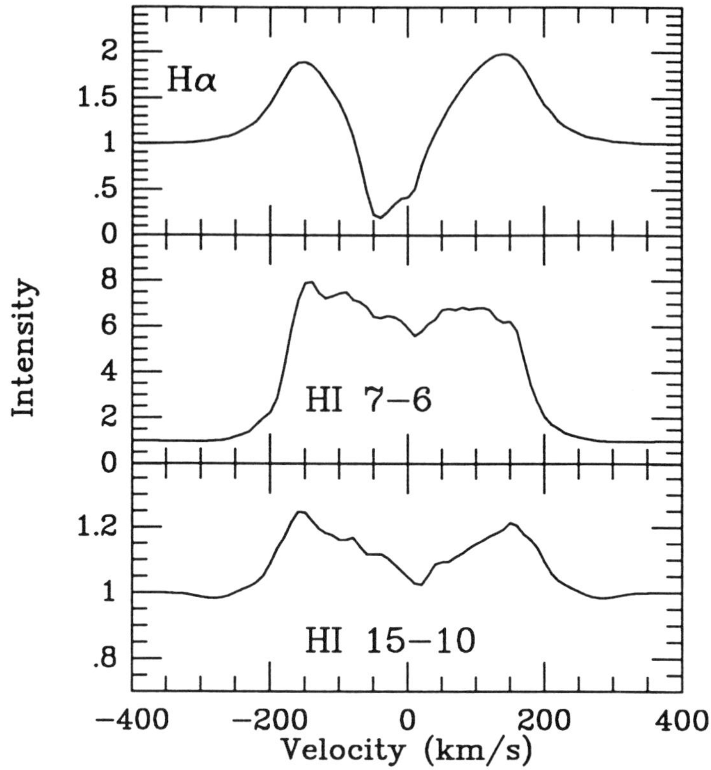

Fig. 5. Examples of theoretical line profiles calculated for a Be star with a rotating and expanding disc of high density gas, seen edge-on. Wavelengths are 0.6562 μm (Hα), 12.37 μm (HI 7-6), and 16.41 μm (HI 15-10). Notice the asymmetry of the lines, which is opposite for the optical and IR lines, and which is due to small radial motions.

shape $v(r) = v_0(r/R_*)^{n-2}$. We used an initial velocity of 1.6 km/s. The rotational velocity was assumed to be 0.8 times break-up velocity, corresponding to 460 km/s, and Keplerian. We show three lines, the optical Hα line, the HI 7-6 (λ 12.37 μm) and the HI 15-10 (λ 16.41 μm) transition. The line shape of these three lines is quite different, and their differences can be explained in terms of the region in which they are formed, in combination with the effects of the continuum opacity on the line shape (especially the IR lines).

All lines in Fig. 5 show a double-peaked structure, which is easily explained in terms of the rotating disc, in which the receding part of the envelope causes the red emission, and the approaching half the blue emission. In the case of no radial motions (very small v_0), the blue and red peaks should be equally strong, but in the presence of (even small) radial motions the lines are not symmetric any more, as is the case in this example. The remarkable effect occurs that this radial motion causes the *red* peak to become stronger in the optical line, while the *blue* peak is stronger in the IR line. This can be understood by noting that the Hα line has a strong central reversal, caused by the material that is projected against the disc of the star, and which absorbs photospheric radiation. This absorption will be centered at zero velocity in the case of no radial motions, but it will be blue-shifted when there is outflow, thus reducing the strength of the blue emission compared to the red one. In the limit of very high radial motions this will result in a P Cygni line profile. The HI 7-6 line however does not show a strong central reversal: the line opacity is much stronger than the continuum opacity, and the stellar photospheric flux is negligible at these wavelengths. Furthermore, the continuum opacity in the wind will result in an obscuration of the far end of the wind, which is on average moving away from the observer. This will result in a blue peak stronger than the red one. Finally, the HI 15-10 line is weak and more or less symmetric, because it is formed close to the star, and its opacity is not too different from that of the adjacent continuum. Close to the star the motions are predominantly rotational, which will result in a symmetric line profile. This example shows how sensitive the lines are to the details of the velocity field in the equatorial regions.

4.3. What Can a Large IR Telescope Do?

Far-IR observations of hot stars could help us to better understand their mass loss and thus their evolution. The continuum energy distribution should be measured with an accuracy of 1 percent between 2 and 100 μm, to measure accurately the onset of the IR excess and its wavelength dependence. At the longer wavelengths, the background from the galactic plane may become a problem, but this could be resolved by the large spatial resolution. The IR recombination lines and fine structure lines should be observed with a resolution of 5 to 50 km/s, in order to determine the velocity law and the mass loss rate (the HI lines) and the chemical composition of the wind and the ejecta (fine structure lines).

There is evidence that the mass loss of O stars in galaxies with low metallicity is smaller than that in the Milky Way (e.g. Hutchings *et al.*, 1987). The high sensitivity of a large cooled telescope, in combination with the spatial resolution, will allow investigations of stars in other galaxies. This will give insight in the mass loss and wind structure of stars with different metallicities, and test theoretical predictions of the effect of metallicity on mass loss and evolution. In Table II we give the expected IR continuum fluxes of the LBV P Cygni (Fig. 1), and scale it to the distance of some nearby galaxies. With the sensitivity of EDISON quoted by Thronson *et al.* (1991), we should be able to detect such stars to a distance of several Mpc.

5. AGB and Post-AGB Stars

All stars with masses below about 8 $M_\odot$ end their life as a white dwarf of order 0.6 - 1 $M_\odot$, and will necessarily have to lose a large fraction of the inital mass during their evolution. Observations indicate that most of this mass is lost after the main sequence, i.e. when the star is a red giant. Indeed, the late stages of evolution of low and intermediate mass stars are characterized by substantial mass loss. This

TABLE II
IR continuum of P Cygni scaled to other galaxies

$\lambda(\mu m)$	P Cygni (Jy)	LMC (mJy)	M31 (mJy)	M81 (mJy)
2.2	37.9	49.2	0.23	0.013
4.8	15.8	20.5	0.097	5.3 (-3)
12	5.7	7.4	0.035	1.9 (-3)
25	3.5	4.5	0.022	1.2 (-3)
60	2.52	3.3	0.016	0.8 (-3)

mass loss to a large degree determines the evolution of stars on the Asymptotic Giant Branch (AGB) and beyond (for a review, see Habing (1989)), and, as is the case for the massive stars, this evolution cannot be understood without accurate knowledge of the mass loss rate along the evolution. To date, no theory from first principles exists that describes mass loss of AGB stars, and the mass loss rates adopted in evolutionary calculations are uncertain.

5.1. The AGB Evolution of Low Mass Stars

AGB stars have an inert degenerate C/O core, surrounded by a hydrogen burning shell, and occasionally by a helium burning shell. The H-shell provides the luminosity of the star most of the time, but when sufficient He is accumulated by the H-burning, this will ignite. This thermal pulse lasts a relatively short period of time, and during such a pulse the He-shell provides the luminosity. The pulses also cause a large amount of mixing of nuclear processed material into the upper atmosphere, causing the carbon abundance to be enhanced (third dredge-up). If sufficient C is mixed into the atmosphere, the star will become a carbon star. About 20 percent of all AGB stars are C-rich.

The evolutionary timescale of AGB stars (and also of post-AGB stars) is determined by the speed at which the envelope is removed. This in turn depends on the nuclear burning rate (and thus luminosity) of the star, and it depends on the mass loss rate. The observations of AGB stars indicate that during this relatively short evolutionary phase, lasting perhaps as short as 10^4 years, the mass loss is extremely high (typically 10^{-5} $M_\odot$/yr and as high as 10^{-4} $M_\odot$/yr), which means that the mass loss governs the evolution of the star!

The massive, dense winds of AGB stars are characterized by low expansion velocities, of the order of 10-20 km/s, and dust can condense in the outflow. This dust will absorb the optical and UV radiation of the central star, and will re-emit it in the IR at some equilibrium temperature. If the mass loss is high enough, the star will not be detectable in the optical, but will emit all flux in the (far)-IR: the OH/IR stars. The IRAS observations of AGB stars indeed indicate that these stars have colour temperatures well below 1000 K. In Figure 6 we show the IRAS far-IR

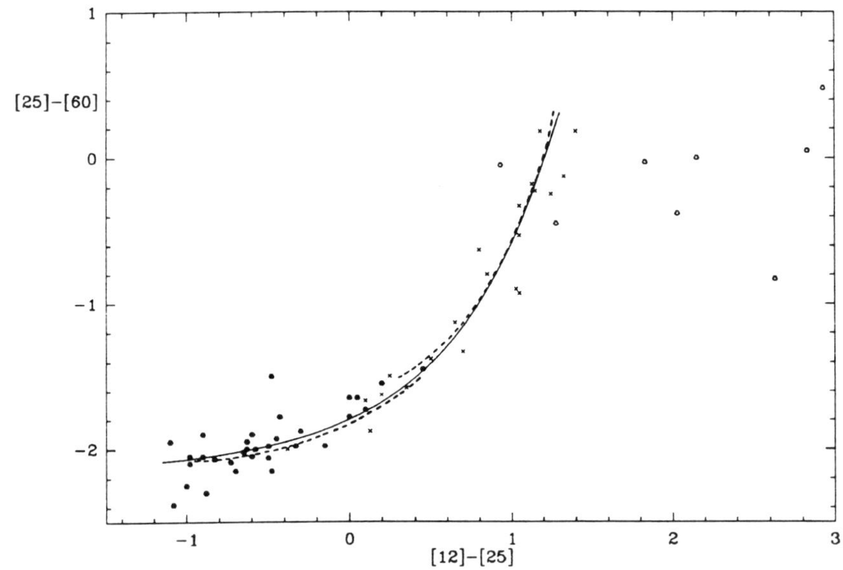

Fig. 6. IRAS far-IR colour-colour diagram of cool AGB stars (Miras and OH/IR stars), taken from van der Veen and Habing (1988). The dashed lines are model calculations by Bedijn (1987) for AGB stars with increasing mass loss, the solid line is a fit to the observed distribution of stars. Mass loss rates can be up to 10^{-4} $M_{\odot}$/yr.

colour-colour diagram of AGB stars (taken from van der Veen and Habing (1988)). The stars form a sequence of increasing redness, which can be interpreted as an increase in the mass loss rate. The reddest stars in this diagram have mass loss rates above 10^{-4} $M_{\odot}$/yr.

In order to understand the evolution of AGB stars, it is necessary to know the change of mass loss as a function of time. The distribution of AGB stars in the IRAS colour-colour diagram has been interpreted by Bedijn (1987) as an evolutionary track, i.e. stars start off with rather low mass loss, and evolve towards very high mass loss rates, thereby constantly reducing the IR colour temperature. However, this interpretation meets with some difficulties, and an alternative interpretation could be that the sequence of increasing redness is a mass sequence, i.e. the more massive stars being those with the highest mass loss and thus the reddest.

At some point the entire envelope of the star has been removed, and the mass loss must drop to much lower values. The AGB shell will begin to expand and cool, and after some time (probably only several 100 years, depending on the mass in the AGB shell) it will become optically thin. The central star will become visible, and the object is recognizable as a post-AGB star. The star will continue to move to higher effective temperatures due to the further reduction of the envelope mass as a result of both nuclear burning and post-AGB mass loss. When the star reaches a temperature of about 2 10^4 K, it will begin to ionize the AGB remnant, which is observable as a planetary nebula (PN).

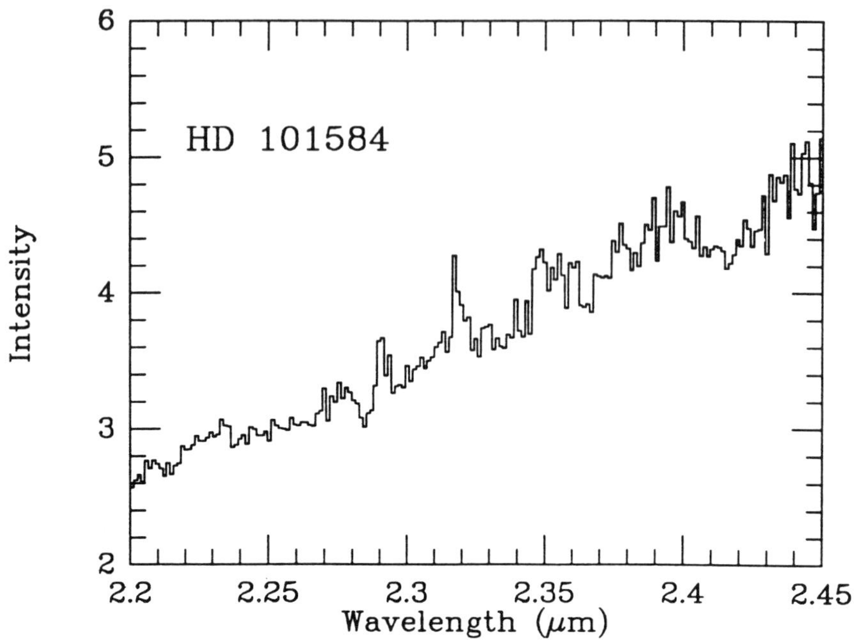

Fig. 7. K-band spectrum of the probable post-AGB star HD 101584, taken with the 3.6 m telescope at ESO. The emission at 2.29 and 2.32 μm is due to CO, indicating a post-AGB molecular outflow in this object.

The short evolutionary phase between the AGB and the PN stages is important for our understanding of the formation of planetary nebulae. For instance, mass loss on the AGB is mostly spherically symmetric, while the majority of PNe have strong deviations from sphericity. The origin of the non-sphericity could be in the post-AGB phase, but it is not clear what physical process could be responsible. It is striking to note that many of the known post-AGB objects have a bipolar morphology. Post-AGB mass loss can be very high in some cases (e.g. Trams *et al.*, 1990; Slijkhuis *et al.*, 1991), which will speed up the removal of the remaining atmosphere, and thus also the evolution towards higher effective temperatures. Again, this has quite an impact on the formation of planetary nebulae. Spectroscopy of lines of ions and molecules can be used to obtain information on the post-AGB wind (outflow velocity, mass loss rates). In Figure 7 we show the K band spectrum of the probable post-AGB star HD 101584 (F2 I), taken by W. van der Veen with the IRSPEC spectrograph attached to the 3.6 m telescope at ESO. The emission at 2.29 and 2.32 μm is due to CO, but unfortunately the resolution is not sufficient to derive the outflow velocity of the molecular gas.

5.2. What Can a Large IR Telescope Do?

It is clear that the study of AGB and post-AGB stars requires observations in the (far)-IR wavelength region. The high spatial resolution of a large IR telescope will

TABLE III
Spatial extent of a typical AGB wind (in arcseconds)

$T_{dust}(K)$	100 pc	200	300	500	1000
1000	0.27	0.13	0.09	0.05	0.03
500	1.47	0.73	0.49	0.29	0.15
300	5.47	2.73	1.82	1.09	0.55
200	14.7	7.33	4.89	2.93	1.47
100	80.0	40.0	26.7	16	8

allow detailed investigations of the spatial structure of AGB and post-AGB stars, as well as PNe. In Table III we list the expected size in arcsec of an envelope around a typical AGB star with $R_* = 600\ R_\odot$, and $T_* = 2500$ K (corresponding to $L = 12000\ L_\odot$). For sake of simplicity we assumed the dust is optically thin, which will slightly overestimate the size of the envelope if the optical depth is substantial. The table demonstrates that the envelopes of nearby AGB stars should be resolvable in the 10-100 μm range. This allows an independent determination of the temperature distribution of the dust in the envelope, and puts much more constraints on the dust opacity and the density gradient in the wind.

High spatial resolution will also give insight into the geometry of the AGB and post-AGB winds, in particular on the onset of non-spherical mass loss. This may occur at the very end of the AGB, or just after the star evolved off the AGB. It also allows a study of the effect of binarity on AGB evolution, of which little is known at present.

Spectral imaging at moderate spectral resolution ($R \approx 500$) may reveal changes in the chemical composition of the dust as function of distance from the central object. As mentioned in the previous Subsection, a significant fraction of AGB stars change their surface chemical abundances from O-rich to C-rich due to mixing during thermal pulses. This results in a change in chemistry of the circumstellar dust as well. A measurement of such a change in dust chemistry would be direct evidence that a star has recently changed from O-rich to C-rich, and would furthermore allow us to derive the time at which this happened.

Very high spectral resolution observations ($R \approx 10^6$) of the OH absorption at 35 μm would allow detailed modeling of the OH maser pump mechanism. It is presently assumed that the 35 μm transition is the main pump for the OH maser at 1612 MHz. The shape of the 35 μm line is sensitive to the velocity field in the OH shell, and also to the location of the OH with respect to the region where the 35 μm continuum photons are generated.

The high sensitivity will allow measurements of the energy distribution of AGB stars in external galaxies. This will give information on the effect of metallicity on the formation of dust, and on the mass loss rates. It would also allow the study of the dust features in individual stars in other galaxies.

Spectral imaging in the far-IR of PNe would give a better insight into the distribution of ions of metals as Fe, Si and Mg. There are no useful fine structure lines of these elements in the optical, and at present the abundances of metals in PNe are inferred indirectly. It is thought that Fe, Si and Mg are strongly depleted because it is locked up in grains. However, when the grains are destroyed by the intense UV radiation field of the hot central star, they should be returned to the gas, and should be measurable in the innermost part of the nebula.

6. Conclusions

The infrared wavelength region provides a unique opportunity to study the circumstellar material in various kinds of stars, ranging from Vega-type stars to AGB stars and planetary nebulae. Both the continuum and the spectral features give excellent diagnostic tools for the analysis of the physical conditions in the stellar winds of stars, and allow us to study mass loss and its effect on evolution of stars of a wide range in initial masses. A large space-borne IR telescope, such as EDISON, could significantly contribute to our knowledge.

In order to further build on the knowledge that ISO and SIRTF will hopefully provide, the focal plane instruments should be capable of spectral resolutions of the order of 10^5 to 10^6, while also spectral imaging at a moderate spectral resolution (R = 1000) would be extremely valuable for various kinds of studies. The wavelength coverage should be as broad as possible, preferentially extending to 100 μm, because especially the longer wavelengths are not accessible from the ground, and even with the degraded performance of a mechanically cooled telescope at those longer wavelengths, there is still a lot of exciting science that can be done.

Acknowledgements

I thank Bert Zijlstra and Norman Trams for useful discussions, and Mike Marlborough for providing models for Be star recombination lines. Wil van der Veen is thanked for use of the K band spectrum of HD 101584 prior to publication. B. Gustafsson is kindly acknowledged for providing a model atmosphere for ϵ Eri.

References

Aumann, H. H., *et al.*: 1984, *Astrophys. J.* **278**, L23
Aumann, H. H.: 1985, *Publ. Astron. Soc. Pacific* **97**, 885
Barlow, M. J., Roche, P. F., Aitken, D. K.: 1988, *Monthly Notices Roy. Astron. Soc.* **232**, 821
Bedijn, P.: 1987, *Astron. Astrophys.* **186**, 136
Bertout, C., Leitherer, C., Stahl, O., Wolf, B.: 1985, *Astron. Astrophys.* **144**, 87
Bieging, J. H., Abbott, D. C., Churchwell, E. B.: 1989, *Astrophys. J.* **340**, 518
Bruhweiler, F. C., Kondo, Y., Grady, C. A.: 1991, *Astrophys. J.* **371**, L27
Cassinelli, J. P., Swank, J. H.: 1983, *Astrophys. J.* **271**, 681
Castor, J. I.: 1970, *Monthly Notices Roy. Astron. Soc.* **149**, 111
Castor, J. I., Abbott, D. C., Klein, R. I.: 1975, *Astrophys. J.* **195**, 157

Chu, Y. Y.: 1991, in K. A. van der Hucht and B. Hidayat, ed(s)., *IAU Symp. 143: Wolf-Rayet stars and interrelations with other massive stars in Galaxies*, Reidel, 349

Gillett, F. C.: 1986, in F. P. Israel, ed(s)., *Light on Dark Matter*, D. Reidel Publ. Company, Dordrecht, the Netherlands, 61

Groenewegen, M. A. T., Lamers, H. J. G. L. M.: 1989, *Astron. Astrophys.* **79**, 359

Habing, H. J.: 1989, in M. O. Mennessier and A. Omont, ed(s)., *From Mira's to Planetary Nebulae: which path for stellar evolution?*, Edition Frontières, 16

Harvey, P. M., Wilking, B. A., Joy, M.: 1984, *Nature* **307**, 441

Humphreys, R. M., Davidson, K.: 1979, *Astrophys. J.* **23**, 409

Hutchings, J., Massey, P., Bianchi, L.: 1987, *Astrophys. J.* **322**, L79

Kurucz, R. L.: 1979, *Astrophys. J. Suppl.* **40**, 1

Lagrange-Henri, A. M., Beust, H., Ferlet, R., Vidal-Madjar, A.: 1989, *Astron. Astrophys.* **215**, L5

Lamers, H. J. G. L. M., Waters, L. B. F. M.: 1984, *Astron. Astrophys.* **136**, 37

Lamers, H. J. G. L. M., Waters, L. B. F. M., Garmany, C. D., Perez, M. R., Waelkens, C.: 1986, *Astron. Astrophys.* **154**, L20

Maeder, A., Meynet, G.: 1987, *Astron. Astrophys.* **182**, 243

Nota, A. and Paresce, F.: 1989, in K. Davidson, A. F. J. Moffat and H. J. G. L. M. Lamers, ed(s)., *Physics of Luminous Blue variables*, Kluwer Academic Publishers, Dordrecht, the Netherlands, 159

Roche, P. F.: 1989, in , ed(s)., *Infrared Spectroscopy in Astronomy*, ESA SP-290, 79

Simon, M., Felli, M., Cassa, L., Fischer, J., Massi, M.: 1983, *Astrophys. J.* **266**, 623

Slijkhuis, S., Hu, J. Y., de Jong, T.: 1991, *Astron. Astrophys.* , in press

Sobolev, V. V.: , in V. A. Ambartsumyan, ed(s)., *Theoretical Astrophysics*, , Chapter 29

Thronson, H. A., Hawarden, T. G., Mountain, C. M., Davies, J. K., Lee, T. J., Longair, M.: 1991, in Y. Komoto, ed(s)., *Observatories in Earth orbit and beyond, IAU colloquium 123*, Kluwer Academic Publishers, Dordrecht, the Netherlands,

Trams, N. R., van der Veen, W. E. C. J., Waelkens, C., Waters, L. B. F. M., Lamers, H. J. G. L. M.: 1990, *Astron. Astrophys.* **233**, 153

van der Veen, W. E. C. J., Habing, H. J.: 1988, *Astron. Astrophys.* **194**, 125

Walker, H. J., Wolstencroft, R. D.: 1988, *Publ. Astron. Soc. Pacific* **100**, 1509

Waters, L. B. F. M., Wesselius, P. R.: 1986, *Astron. Astrophys.* **155**, 104

Waters, L. B. F. M., Coté, J., Geballe, T. R.: 1987, *Astron. Astrophys.* **203**, 348

THE INFRARED SPACE OBSERVATORY (ISO)

M.F. KESSLER, L. METCALFE and A. SALAMA
ISO Science Operations Team, Astrophysics Division, European Space Agency, Noordwijk, The Netherlands

Abstract. The Infrared Space Observatory (ISO), a fully approved and funded project of ESA, will operate at wavelengths from 3–200 microns. The satellite essentially consists of a large cryostat containing about 2300 litres of superfluid helium to maintain the telescope (primary mirror diameter of 60 cm) and the scientific instruments at temperatures between 2K and 8K. A pointing accuracy of a few arc seconds is provided by a three-axis-stabilisation system. ISO's instrument complement consists of four instruments, namely: an imaging photo-polarimeter (3–200 microns), a camera (3–17 microns), a short wavelength spectrometer (3–45 microns) and a long wavelength spectrometer (45–180 microns). ISO's scheduled launch date is May 1993 and it will be operational for at least 18 months. In keeping with ISO's role as an observatory, two-thirds of its observing time will be made available to the general astronomical community via several *Calls for Observing Proposals*.

1. Introduction

The scientific potential of infrared astronomy has already been amply demonstrated by observations with ground-based telescopes using infrared instruments operating within the few atmospheric windows, as well as by observations with aircraft- and balloon-borne telescopes. However, all these systems are limited by atmospheric absorption and also by the photon shot-noise associated with the thermal emission from the warm telescope optics and from the atmosphere. Operating outside the Earth's atmosphere with a cooled telescope and instruments removes these limitations and allows a major increase in sensitivity.

The first major step in this direction was taken with the Infrared Astronomical Satellite (IRAS), which surveyed nearly all of the sky in four broad infrared bands between 8 and 120 microns. IRAS was so successful, that even now, 8 years after it finished its in-orbit mission, exciting new results are still appearing in the literature. ISO, being an observatory satellite, will build on the results of IRAS and the new generation of sub-mm telescopes by making detailed observations of selected sources. Compared to IRAS, ISO will have a longer operational lifetime, wider wavelength coverage, better angular resolution, more sophisticated instruments and, through a combination of detector improvements and longer integration times, a sensitivity gain of up to several orders of magnitude.

ISO will provide astronomers with a facility of unprecedented sensitivity for detailed exploration of the Universe, ranging from objects in the solar system right out to the most distant extragalactic sources. Its cryogenically-cooled telescope will be equipped with four scientific instruments, which together will permit imaging, and photometric, spectroscopic and polarimetric observations at wavelengths from 2.5–200 micron.

Space Science Reviews **61:** 45–60, 1992.

ISO is designed to be a true observatory, with its scientific instrumentation capable of tackling a wide range of astrophysical problems and with two-thirds of its observing time available to the general community.

2. Scientific Rationale

The ISO part of the spectrum is of great scientific interest, not only because it is here that cool objects (15–300K) radiate the bulk of their energy, but also because of its rich variety of atomic, ionic, molecular and solid-state spectral features. Measurements at these wavelengths permit determinations of many physical parameters; for example, energy balance, temperatures, abundances, densities and velocities. Owing to the much-reduced extinction, infrared observations are particularly well-suited to probing the properties of objects obscured at visible wavelengths. As ISO will be offering high-sensitivity and sophisticated observing facilities for a relatively unexplored part of the spectrum, it is expected that the ISO scientific programme will have an impact on virtually every field of astronomy.

3. Spacecraft

The complete satellite, shown in Figure 1, is 5.3 m high, 2.3 m wide and will weigh around 2400 kg at launch. It is divided into a payload module and a service module. These modules are built up of various sub-systems, which are described in the next two sections.

3.1. Payload Module

The payload module is essentially a large cryostat containing the telescope and the scientific instruments. The main sub-systems of this module are:

Helium Sub-system: This sub-system provides the cooling power to maintain, *inter alia*, the scientific instruments and telescope at temperatures between 1.8 and 4 K for an operational lifetime in-orbit of at least 18 months. The main toroidal HeII tank, equipped with a porous-plug phase separator, contains 2250 litres of superfluid helium at a temperature of 1.8 K. Some of the infrared detectors are directly coupled to this helium tank; all other units are cooled using the cold boil-off gas from the liquid helium. This gas is first routed through the optical support structure to cool the telescope and the scientific instruments. It is then passed along the optical baffles and radiation shields before being vented to space. A small auxiliary tank, containing about 60 litres of normal liquid helium, fulfills all of ISO's cooling needs for the last 72 hours before launch.

Optical Sub-system: The main components of this sub-system are the telescope, the set of baffles and the sunshade. Suspended in the middle of the main helium tank is the telescope, which has an aplanatic Ritchey-Chrétien configuration

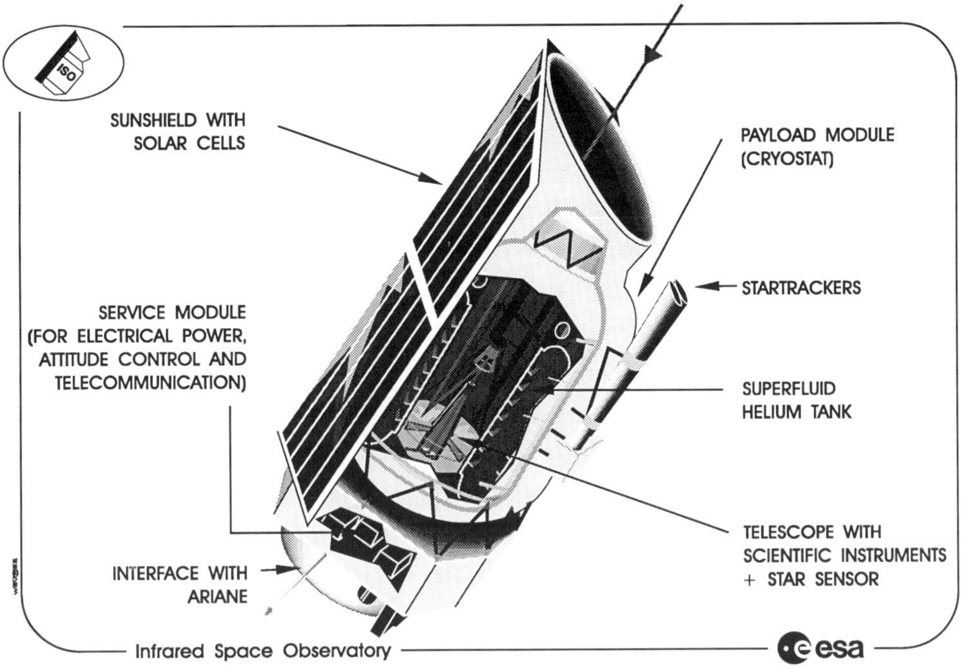

Fig. 1. A computer graphic of the ISO satellite, showing the Payload and Service Modules in cutaway view.

with an effective aperture of 600 mm and an overall f/ratio of 15. A four-sided pyramid mirror distributes the 20′ unvignetted field of view of the telescope among the four instruments, which are mounted on the side of the optical support structure opposite to the primary mirror. Each instrument simultaneously receives a 3′ unvignetted field centred on an axis at an angle of 8.5′ to the telescope optical axis. A weight-relieved fused-silica (Herasil I) primary mirror and a solid fused-silica secondary mirror have been selected as the telescope optics. The optical quality of these mirrors is adequate for diffraction-limited performance at a wavelength of 5 micron.

Stringent control of stray light, particularly from bright infrared sources outside the telescope's field of view, is necessary in order to ensure that the system sensitivity is not degraded. This control is accomplished (i) by imposition of viewing constraints (ii) by equipping the mirrors with baffles (iii) by the main optical baffle, which protects from off-axis straylight and thermal self-emission from the sunshade and (iv) by the sunshade, which prevents earth and sun radiation from entering the telescope.

Structure and Insulation Sub-system: The largest component is the vacuum vessel, which is composed of three forged and machined parts, rigidly flanged to each other. Multi-layer insulation (MLI) is used on the outside of the vacuum vessel. All internal units are suspended from the vacuum vessel via glass-fibre and carbon-fibre straps and spatial frameworks. The three gas-cooled radiation shields are each covered with MLI and there is an additional shield to provide a light-tight enclosure for the instruments. Mounted on the outside of the vacuum vessel is a sunshield, which prevents the sun from shining directly on the cryostat.

Cryostat Cover Sub-system: This cover provides many facilities beyond its basic purpose of closing the cryostat on the ground. It contains two windows which transmit visible radiation, thereby permitting the optical alignment of the telescope and the instruments to be monitored even when they are at cryogenic temperatures. To aid instrument testing, the cover carries 2 infrared sources which can 'illuminate' the instruments and, additionally, the cover's innermost radiation shield can be cooled to below 5 K with a liquid-helium flush so as to provide a low thermal radiation background. The cover is secured by a clampband and it will be ejected about 14 days after launch after the spacecraft has outgassed sufficiently.

3.2. Service Module

As its name indicates, the service module provides the 'classical' support services to the entire satellite. These include:

- the structure and the load path to the launcher,
- the solar array mounted on the sunshield, delivering 580 W at the end of the mission,
- sub-systems for thermal control, data handling, power conditioning, telemetry (nominal rate 32768 bps of which about 24 kbps are available for the scientific instruments) and telecommand (2000 bps),
- the 'attitude and orbit control' subsystem. This provides the three-axis stabilisation and the raster pointing facilities needed for the mission. The requirements on various pointing errors are:
  = relative pointing error (RPE, equivalent to jitter): 2.7$''$ (2 sigma, half cone angle, over a period of 30secs),
  = absolute pointing error (APE, the angular separation between the commanded direction and the instantaneous actual direction): 11.7$''$ (2 sigma, half cone angle).

  Sensors used are: sun acquisition sensors to provide coarse information on the position of the sun; fine sun sensors for accurate sun position information; earth limb sensors, covering both 'forbidden' and 'warning' zones for the spacecraft attitude with respect to the Earth; star trackers as the prime pointing sensors; a quadrant star sensor on the optical axis of the telescope to calibrate the mis-alignment between the star trackers and the telescope axis; and four rate-integrating gyros in an all-skewed configuration. The main actuators are four reaction wheels, also in an all-skewed configuration. These wheels can be unloaded by use of the hydrazine reaction control subsystem, which is also used for orbit acquisition and maintenance.

4. Scientific Instruments

ISO's instrument complement consists of four instruments: an imaging photopolarimeter (ISOPHOT); a camera (ISOCAM); a short wavelength spectrometer (SWS); and a long wavelength spectrometer (LWS). Each instrument is being built by international consortia of institutes using national funding and will be delivered to ESA for in-orbit operations by an ESA team, which will however include some staff supplied by the groups building the instruments. Although being developed separately, the four instruments have been designed to form a complete, complementary and versatile common-user package, in keeping with the observatory nature of ISO.

Only one instrument will be operational at a time; however, when ISOCAM is not the main instrument, it can be operated in a so-called parallel mode, with a reduced telemetry rate, thus gathering additional astronomical data. A partial sky survey at 200 micron wavelength, a region not explored by IRAS, will be possible by operating ISOPHOT during as many satellite slews as possible. These two extra modes will maximize the overall scientific return of the mission; however they are not selectable by the user.

The spectroscopic and photometric capabilities of ISO are shown in Figure 2. Photometry and polarimetry will be possible in broad and narrow spectral bands across ISO's entire wavelength range from 2.5 to 200 microns. A variety of apertures, ranging from 5 to 180 arcseconds, will be selectable out to 103 microns. For spectroscopy, resolving powers ranging from 50 to 20,000 will be available in the wavelength range 2.5 to 180 microns.

Each instrument will observe adjacent areas of the sky, with a maximum unvignetted field of view per instrument of 3 arcminutes. Figure 3 gives details of the individual instrument apertures. ISO will be capable of imaging in broad and narrow spectral bands across the complete wavelength range at spatial resolution ranging from 1.5 arcseconds (at 2.5–17 microns) to 90 arcseconds (at 120–200 microns).

The ISOCAM instrument (Cesarsky *et al.*, 1989) consists of two optical channels, each with a 32 x 32 element detector array (Figure 4). These arrays operate, respectively, in the wavelength ranges 2.5–5.5 micron and 4–17 micron. The short wavelength array uses an InSb detector with a CID readout and the long wavelength detector is made of Si:Ga with a Direct ReadOut (DRO). Only one channel can be operated at a time, by operating a selection wheel, which also carries two internal flux calibrators for flat-fielding. Each channel contains two further selection wheels: one carries various filters (including Circular Variable Filters, with a resolution of 45) and the other one carries lenses for choosing a pixel field of view of 1.5, 3, 6, or 12 arcsecs. Polarizers are mounted on an entrance wheel common

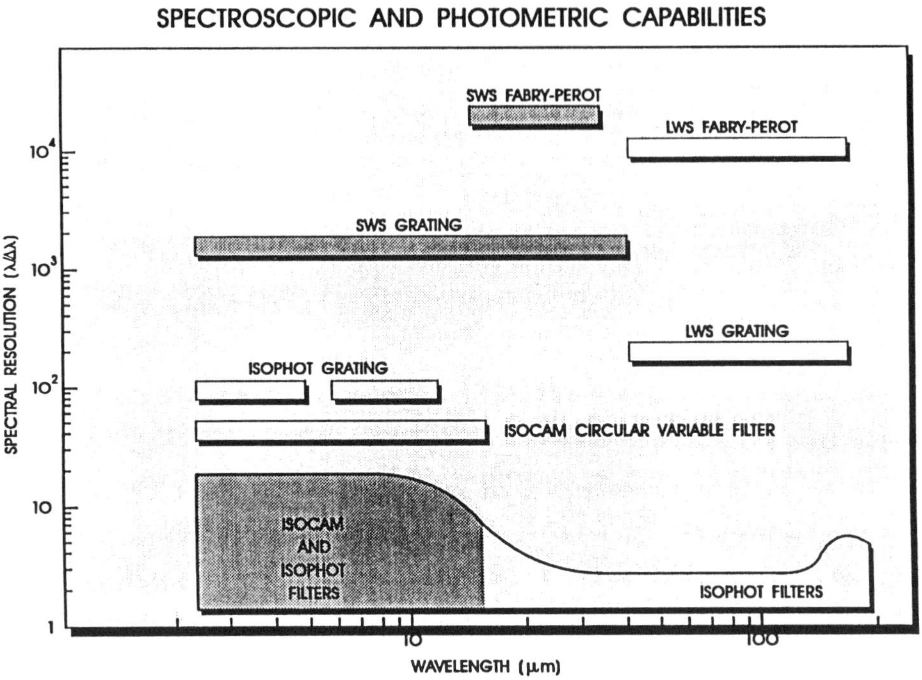

Fig. 2. An overview of the spectroscopic and photometric capabilities of the ISO scientific payload.

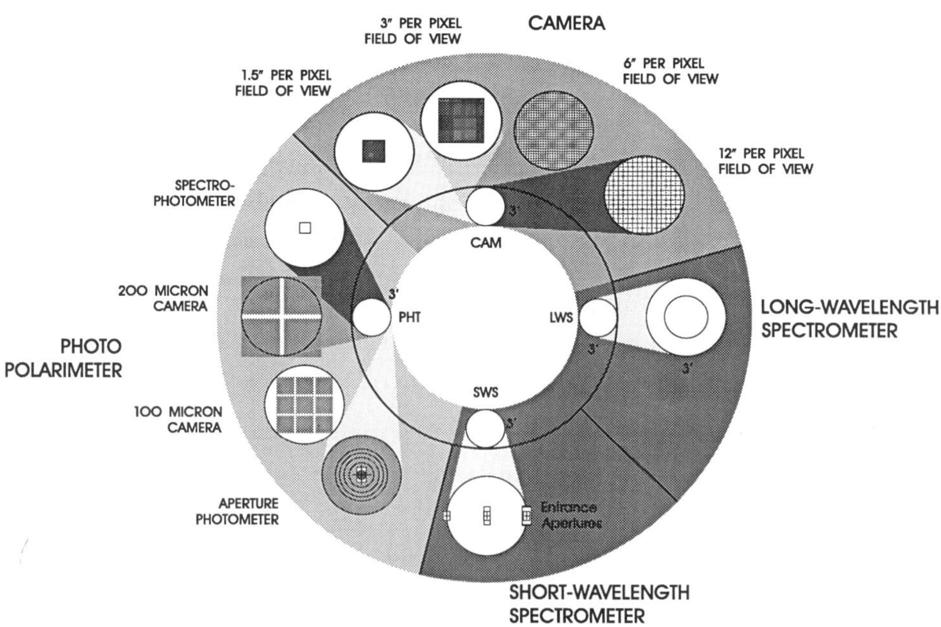

Fig. 3. A schematic view of the fields of view of the ISO instruments as projected onto the sky.

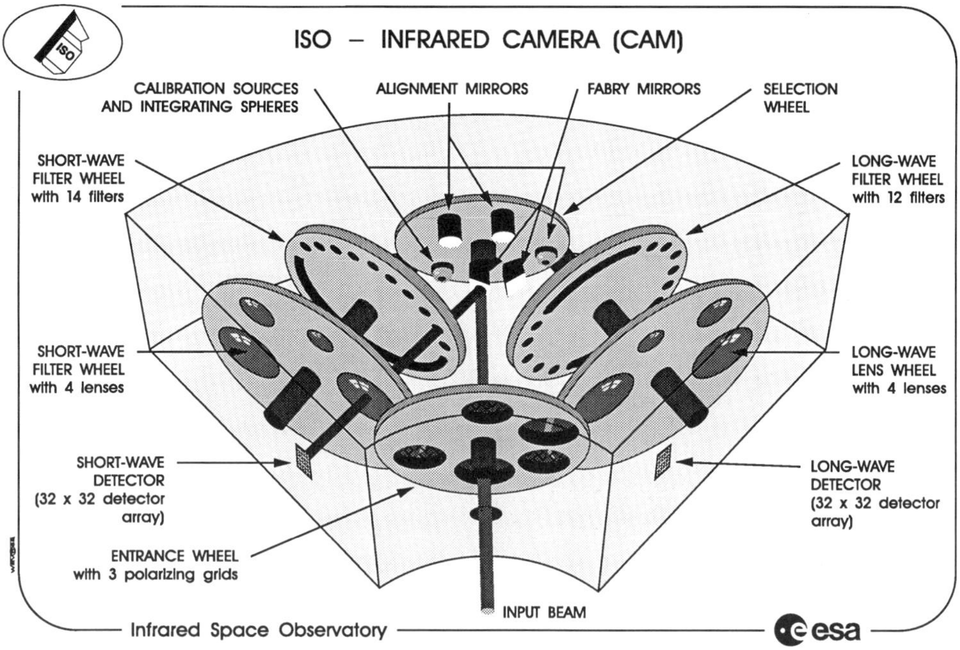

Fig. 4. A schematic view of the ISOCAM instrument (Principal Investigator: C.J. Cesarsky, Saclay, France).

to both channels. Figures 5 and 6 give the predicted sensitivities of ISOCAM for point and extended sources.

The ISOPHOT instrument (Lemke *et al.*, 1989) consists of three subsystems (Figure 7):

- ISOPHOT-C: a photopolarimeter which also provides imaging capability at close to the diffraction limit in the wavelength range from 40 - 200 micron.
- ISOPHOT-P: a multi-band, multi-aperture photopolarimeter for the wavelength range from 30–110 micron.
- ISOPHOT-S: a dual grating spectrophotometer which provides a resolving power of $\sim$ 90 in two wavelength bands simultaneously (2.5–5 micron and 6–12 micron).

A focal plane chopper with a selectable beam throw of up to 3 arcmin is also included in ISOPHOT. Selection between the different modes of the various subsystems is achieved with appropriate settings of three ratchet wheels. ISOPHOT detectors are made of Si:Ga, Si:P BIB, Ge:Ga (stressed and unstressed) and are read-out by specially designed cryogenic integrating amplifier, followed, in most cases, by a multiplexer (CRE). Two redundant sets of thermal radiation sources are located symmetrically about the centre of the ISOPHOT field of view and will be used as a transfer photometric standard to monitor the time evolution of detector responsivity, with flux levels representative of astronomical signals. Figure 8 gives

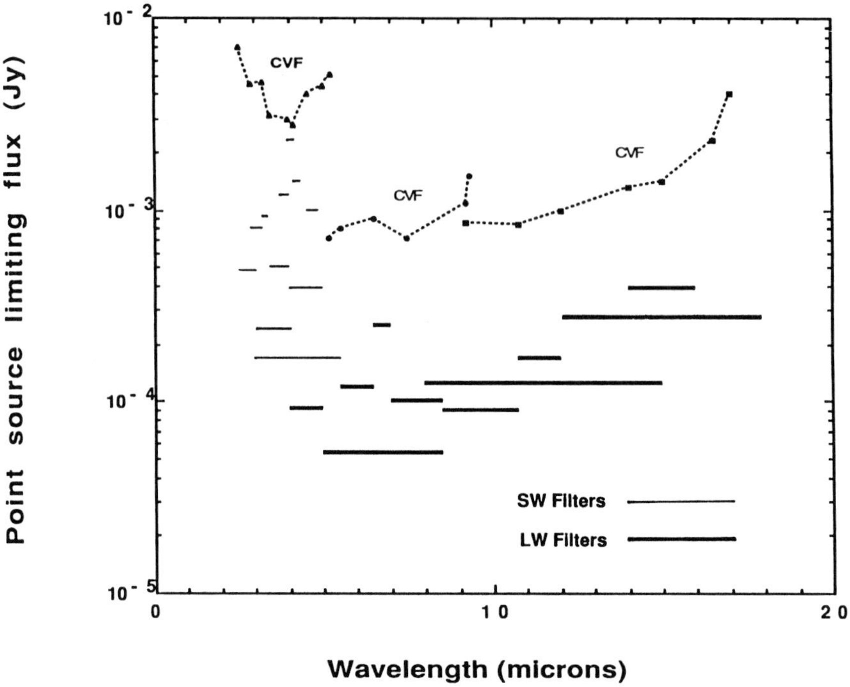

Fig. 5. ISOCAM sensitivity for point sources (10 sigma detection, 3600 secs integration, 6″ pixel field of view).

the predicted sensitivities for ISOPHOT.

The LWS instrument (Emery *et al.*, 1985) consists of a reflection diffraction grating used in 1st and 2nd order with an array of 10 discrete detectors providing a spectral resolving power of around 200 over the wavelength range 45–180 micron (Figure 9). The detectors are made of Ge:Be and Ge:Ga (stressed and unstressed) material and are read-out by JF4 integrating amplifiers. Either of two Fabry-Pérot interferometers can be rotated into the beam by moving an interchange wheel, thereby increasing the resolving power to 10,000 across the entire LWS wavelength range. The single fixed LWS circular field of view is 100 arcsecs, which matches the diffraction limit of the telescope at 118 micron and is a compromise for the wavelength range of the instrument. Five internal illuminators will be used to monitor the detectors' stability. Figures 10 and 11 give the predicted sensitivities of the LWS grating and Fabry-Pérot modes.

The SWS instrument (De Graauw *et al.*, 1989) provides a resolving power of between 1000 and 2000 across the wavelength range from 2.4–45 micron by means of two reflection diffraction gratings used in 1st, 2nd and 3rd orders (Figure 12). Filters for order-sorting are placed at the instrument's three entrance apertures, which can be selected by a suitable pointing of the spacecraft. Projected detector sizes (i.e. pixel sizes), at the entrance apertures, range from $14'' \times 20''$ to $20'' \times 40''\delta$. Over a part of the SWS's operating range, the resolution can be increased to about

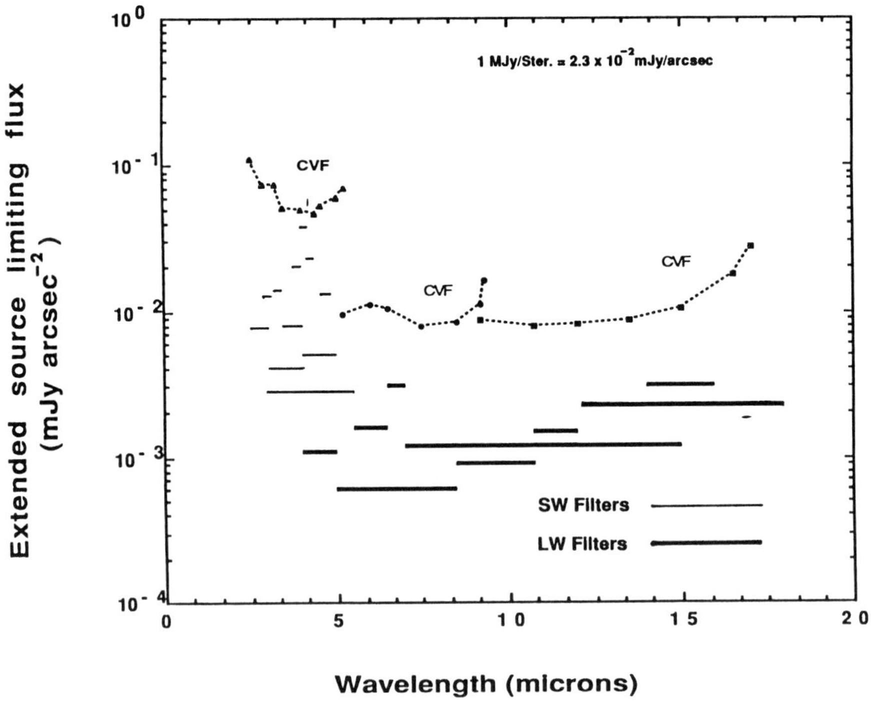

Fig. 6. ISOCAM sensitivity for extended sources (10 sigma detection, 3600 secs integration, 6″ pixel field of view).

20,000 by directing the incident radiation through either of two Fabry-Pérot interferometers. The SWS detectors are made of InSb, Si:Ga, Si:Sb and Ge:Ga material and are read out by IA-12 integrating amplifiers. The SWS beam can be directed to view internal spectral reference sources, located near the instrument entrance and employing fixed Fabry-Pérot resonators for use in wavelength calibrations. IR diffuse illuminators, which will be used to monitor the detectors' stability, are located close to the detectors. Table 1 gives the predicted SWS sensitivity.

5. Orbit and Ground Station Coverage

An Ariane-4 vehicle will launch ISO into a transfer orbit with a perigee height of around 200 km and an apogee height of around 70000 km. ISO's hydrazine reaction control system will then be used to raise the perigee to attain the operational orbit (24-hour period, perigee 1000 km and apogee 70000 km). The inclination to the equator will be around 5°. ESA plans to supply only one ground station, enabling ISO to make astronomical observations during the best 14 hours of each orbit. The mission's scientific return, however, could be greatly increased by the addition of a second ground station, which would permit ISO to be operated for the entire time that it spends outside the main part of the Earth's radiation belts. An international collaboration is being sought by ESA to provide this second ground station.

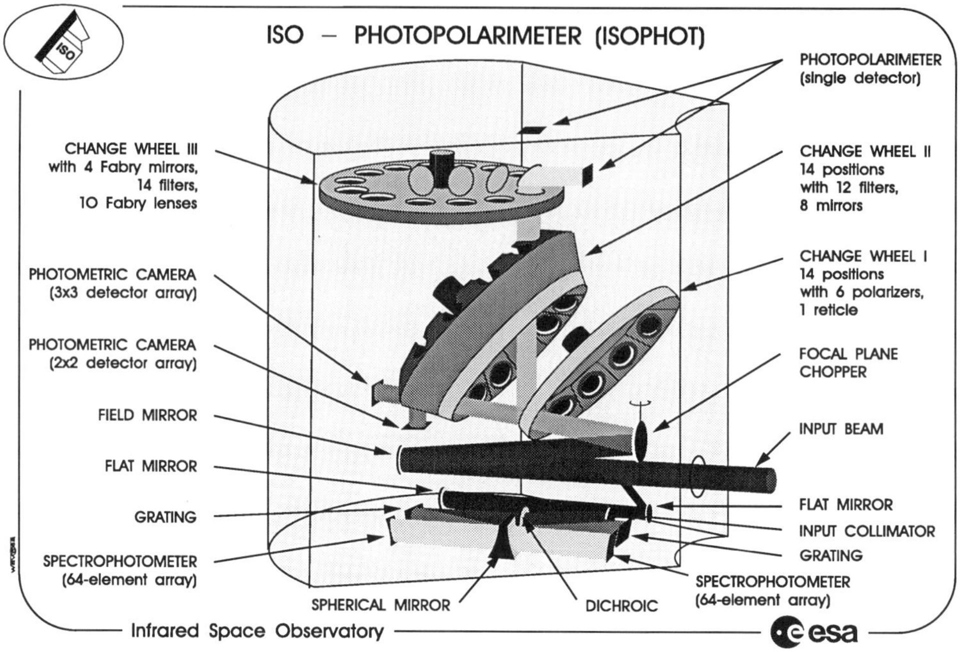

Fig. 7. A schematic view of the ISOPHOT instrument (Principal Investigator: D. Lemke, Heidelberg, Germany).

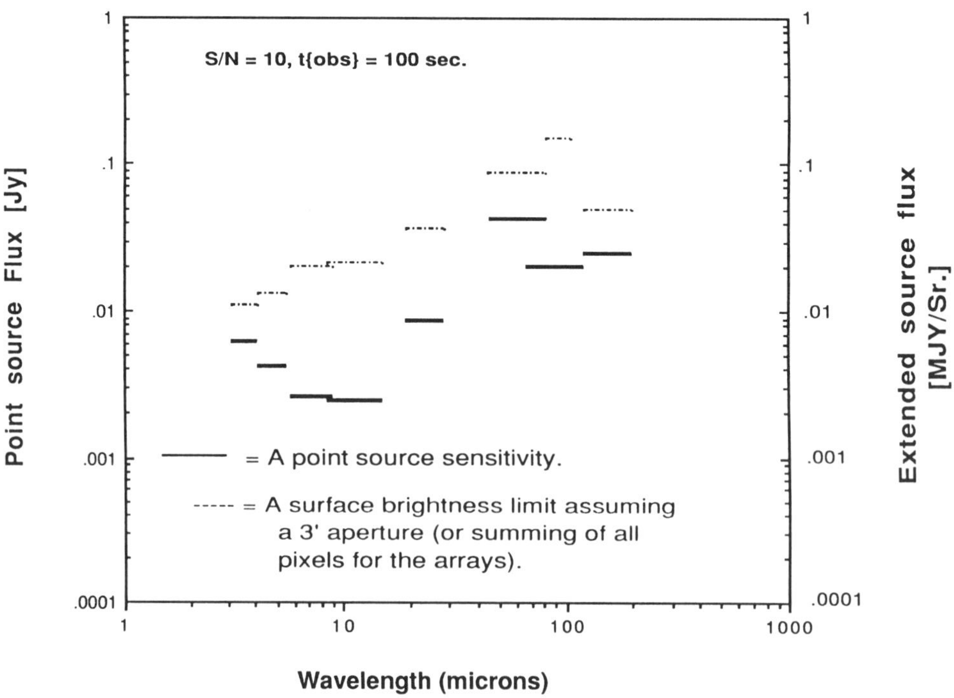

Fig. 8. ISOPHOT sensitivities for broad-band photometry.

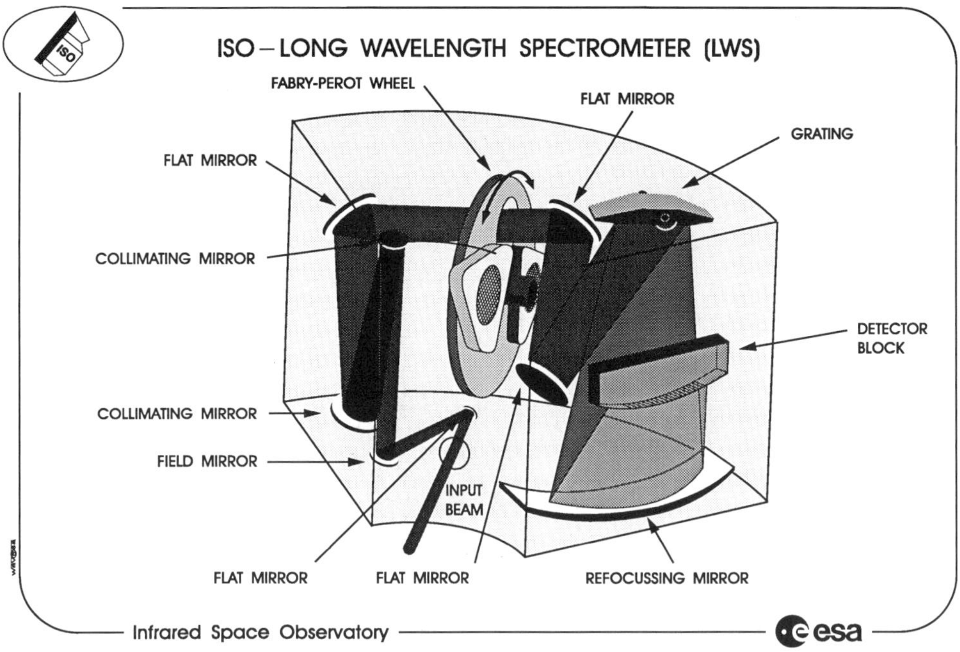

Fig. 9. A schematic view of the LWS instrument (Principal Investigator: P.E. Clegg, London, UK).

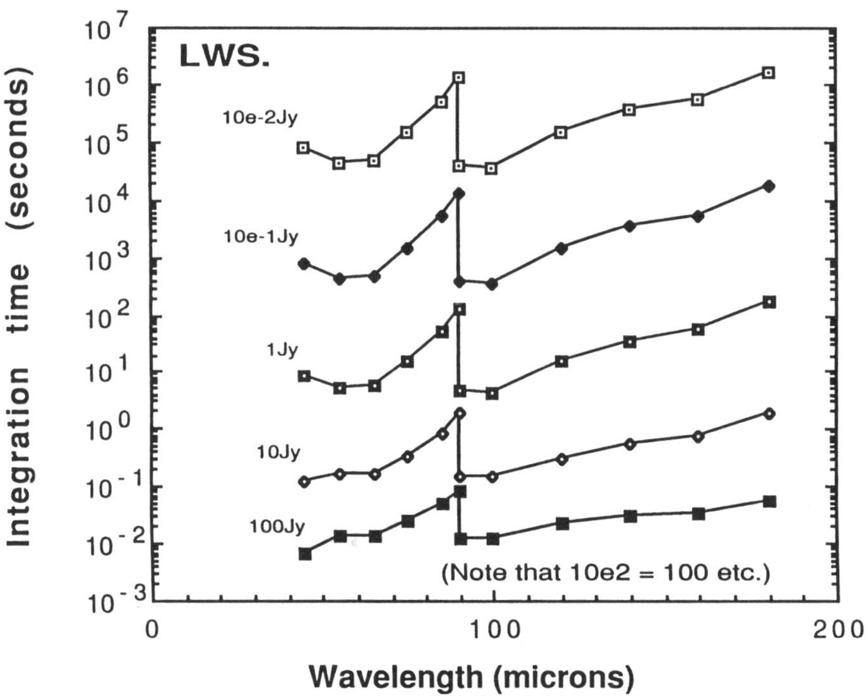

Fig. 10. Integration time required with LWS, in grating mode, to achieve a signal-to-noise ratio of 10, for the spectral flux-densities shown to the left of the curves.

TABLE I

Table I gives the SWS sensitivity to continuum sources, in grating mode. It lists the integration times for 57 wavelength ranges. The upper part of the table refers to the SWS short wavelength section, while the lower part refers to the long wavelength section. The calculations are made for 6 different flux densities. With these integration times, a signal-to-noise ratio of at least 20 is obtained. The table also lists the number of subspectra (N) in each wavelength range (1 subspectrum = the range covered by one detector array with approximately 8 resolution elements) and which entrance aperture of the instrument is used for the subspectra.

N subspectra per λ range				Partial Wavelength Range(s) Grating Scan Observing times for different fluxdensities					
λ_{start} μm	λ_{end} μm	N	Aperture	t (sec.) $S_\nu = 0.3Jy$	t (sec.) $S_\nu = 1Jy$	t (sec.) $S_\nu = 3Jy$	t (sec.) $S_\nu = 10Jy$	t (sec.) $S_\nu = 30Jy$	t (sec.) $S_\nu = 100Jy$
2.430	2.545	10	1	2304	504	168	60	60	60
2.545	2.710	10	1	2304	468	156	60	60	60
2.710	2.873	10	1	1656	360	120	60	60	60
2.873	3.034	10	1	1440	360	120	60	60	60
3.034	3.082	3	1	432	72	36	18	18	18
3.077	3.235	10	2	1440	360	120	60	60	60
3.235	3.407	10	2	2016	360	120	60	60	60
3.407	3.659	10	2	1368	300	120	60	60	60
3.659	3.909	10	2	720	240	90	60	60	60
3.909	4.155	10	2	9360	1032	240	96	60	60
4.155	4.398	10	2	18360	2160	540	120	60	60
4.398	4.693	10	2	22896	2304	720	180	60	60
4.693	4.876	10	2	28224	2880	720	240	60	60
4.876	5.109	10	2	34272	3312	720	240	60	60
5.109	5.533	10	2	23904	2448	576	156	60	60
5.533	6.049	10	2	11880	1440	360	120	60	60
6.049	6.560	10	2	8208	1008	288	120	60	60
6.560	7.066	10	2	5760	720	240	78	60	60
7.066	7.268	4	2	1800	288	96	24	24	24
7.254	7.752	10	3	12456	1440	360	120	60	60
7.752	8.246	10	3	14832	1440	360	120	60	60
8.246	8.734	10	3	17568	2016	504	120	60	60
8.734	9.215	10	3	20376	2160	720	120	60	60
9.215	9.690	10	3	22248	2160	720	156	60	60
9.690	10.16	10	3	25344	2808	720	240	60	60
10.16	10.62	10	3	26496	2880	720	240	60	60
10.62	11.07	10	3	27432	2880	720	240	60	60
11.07	11.52	10	3	28944	2880	720	240	60	60
11.52	11.95	10	3	27792	2880	720	240	60	60
11.95	12.38	10	3	27072	2880	720	240	60	60
12.38	12.80	10	3	25632	2808	720	240	60	60
12.80	12.84	1	3	2376	216	72	24	60	6
12.83	13.65	10	1	447768	40608	4608	720	240	60
13.65	14.47	10	1	346104	31536	4032	720	240	60
14.47	15.27	10	1	269208	24552	3096	720	240	60
15.25	16.04	10	2	76608	7344	1296	240	120	60
16.04	16.82	10	2	98136	9216	1440	360	120	60
16.82	17.59	10	2	120960	11304	1440	360	120	60
17.59	18.35	10	2	150696	13896	2088	360	120	60
18.35	19.09	10	2	185976	17064	2160	396	132	60
19.09	19.82	10	2	227592	20808	2664	720	240	60
19.82	20.53	10	2	280512	25560	3096	720	240	60
20.53	21.41	10	2	277920	25272	3072	600	204	60
21.41	23.10	10	2	9288	1224	324	120	60	60
23.10	24.78	10	2	14256	1512	360	120	60	60
24.78	26.44	10	2	20592	2160	648	144	60	60
26.44	28.07	10	2	29088	3024	720	240	60	60
28.07	28.58	3	2	10512	1080	216	72	18	18
28.47	30.44	10	3	131688	12240	1440	360	120	60
30.44	32.37	10	3	158472	14544	2088	360	120	60
32.37	34.27	10	3	185400	17064	2160	360	120	60
34.27	36.13	10	3	205776	19008	2160	612	204	60
36.13	37.95	10	3	227232	20736	2880	720	240	60
37.95	39.73	10	3	234288	21384	2880	720	240	60
39.73	41.46	10	3	229752	21024	2880	720	240	60
41.46	43.16	10	3	212616	19440	2376	648	216	60
43.16	44.15	6	3	116640	10584	1296	288	96	36

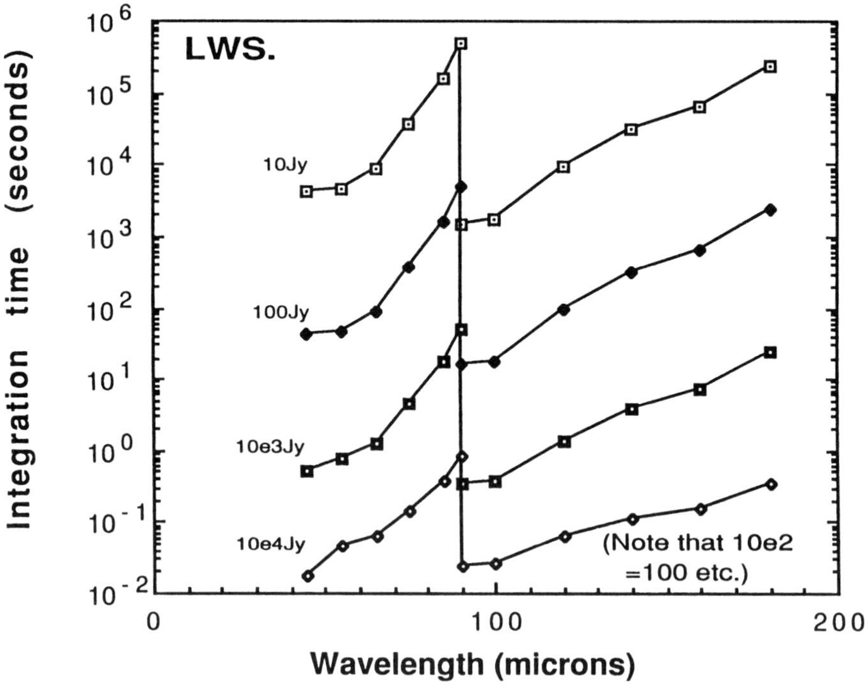

Fig. 11. Integration time required with LWS, in Fabry-Pérot mode, to achieve a signal-to-noise ratio of 10, for the spectral flux-densities shown to the left of the curves.

6. Observing Time and Scientific Operations

Once in orbit, ISO will be operated from ESA's Villafranca satellite control centre (VILSPA), located near Madrid in Spain and currently the home of the IUE observatory. The ISO Science Operations Team will be responsible for all aspects of the scientific operations of ISO ranging from the issue of the 'Calls for Observing Proposals', through the scheduling and use of the scientific instruments, to the reduction and calibration of data for distribution to the observers.

Nearly two-thirds of ISO's observing time will be available to the scientific community via the traditional route of proposal submission, review and selection. In addition to this *Open Time*, there will also be *Guaranteed Time* for the groups who provide the instruments, for the Mission Scientists and for the Science Operations Team. The division of time between these two categories will vary as the mission progresses. After launch, it is anticipated that there will be a period of up to 8 weeks during which the operational orbit will be attained, the spacecraft sub-systems switched on and checked out and the scientific performance of the instruments established. Following this, there will be a 1-month period, consisting of 50% *open* time and 50% *guaranteed* time, during which astronomical observations, designated by the *Observing Time Allocation Committee* as being of the highest priority, will be carried out. For the rest of the mission (at least 15 months), 65%

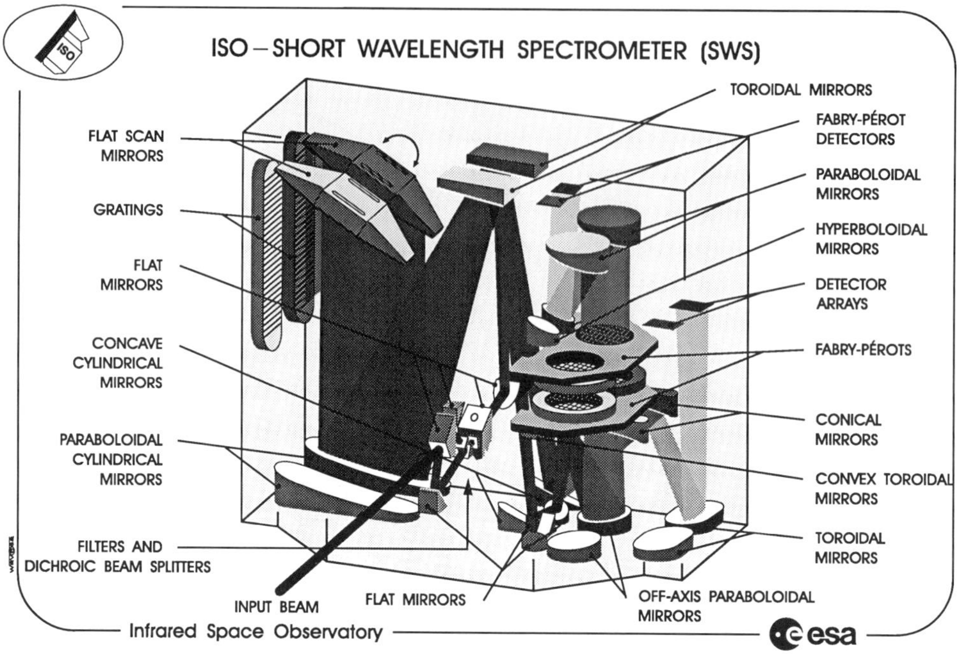

Fig. 12. A schematic view of the SWS instrument (Principal Investigator: Th. de Graauw, Groningen, Netherlands).

of the time will be open time.

A coordinated programme of observations, to be carried out in the guaranteed time, is under preparation by the holders of the guaranteed time. These observations will be published with the Call for Observing Proposals and will be reserved.

The first *Call for Observing Proposals* will be issued between 12 and 18 months before launch. It will solicit proposals for observations to be carried out in the period from 3 to 10 months after launch. During its 18-month lifetime, ISO will be required to make a large number of observations. Thus, a system is being built to automate as much as possible of the process from proposal submission to sending specific commands to the ISO satellite to carry out a particular observation. Proposals will have to be submitted electronically. ESA intends to issue a software package — to run on the proposers' own computers — to help the community prepare these electronic proposals and to check them before transmission to the Science Operations Centre, either via network file transfer or on floppy disks. This package, called the *Proposal Generation Aids,* will contain full details of the predicted performance of the satellite and its instruments as well as electronic 'forms' or 'templates' to be filled in by the proposer to specify **fully** the **exact** details of the observations requested. A second part of the overall software system, called *Proposal Handling,* will be used to check the incoming proposals and extract information for the use of referees and the time allocation committee. The *Observing*

Time Allocation Committee (OTAC) will recommend, via allocation of priorities, which observations should be included in the ISO observing programme. Recommendations will be made on a 'per object' basis, as was the case for EXOSAT, rather than on a 'per shift' or 'per night' basis as is done with IUE and most ground-based telescopes. Based on the grade assigned by OTAC, the *Mission Planning* system will schedule the recommended observations and will generate the detailed instrument commands and spacecraft pointing requests needed for each observation. In order that the best use can be made of the mission's limited lifetime, there will be a review of the implementation of the observing programme about 5 months after launch; if actual spacecraft, instrument or ground segment performances differ from those predicted, the Observing Time Allocation Committee will recommend suitable adjustments to the programme. During the in-orbit phase of the mission, it is planned to issue two further Calls for Proposals so as to give the community the opportunity to react to initial results.

During scientific use, the satellite will always be in real-time contact with the ground segment; however, ISO will be operated according to a detailed, pre-planned schedule in order to maximise the overall efficiency of the mission. Some flexibility in operations will be retained. Examination of the scientific data will be carried out both on- and off-line. In close to real-time, a 'quick-look' output, adequate for an initial estimate of the success or failure of an observation will be available to the Resident Astronomer and Guest Observer (if present). Within a week of an observation being completed, observers will be supplied with a standard set of products from which they may make their astronomical analyses. Per observation, astronomers will receive a standard set of products (in FITS format) including reformatted raw data, reduced images and spectra, calibration data and some specific extracted scientific results to give an indication of what will be found when the observers make their own detailed analysis.

7. Current Status

The initial hardware models of the Payload and Service Modules have been built and successfully tested separately. They will be integrated together during summer 1991 for system-level testing of a development model satellite in the autumn. The flight models of the four scientific instruments are being integrated for calibration and test during summer/autumn 1991 prior to delivery to ESA towards the end of the year. The foreseen launch date is May 1993.

References

Cesarsky, C., Sibille, F. and Vigroux, L.: 1989, *'ISOCAM, a Camera for the Infrared Space Observatory'*, Proc. SPIE **1130**, pp. 202–213.

Emery, R.J., *et al.*: 1985, *'The Long Wavelength Spectrometer (LWS) for ISO'*, Proc. SPIE **589**, pp. 194–200.

Graauw, Th. de *et al.*: 1989, *'The ISO Short Wavelength Spectrometer'*, Proc. 22nd ESLAB Symposium on Infrared Spectroscopy in Astronomy, ESA SP-290, pp. 549–551.

Lemke, D., Burgdorf, M., Hajduk, Ch., and Wolf, J.: 1989, *'Detectors and Arrays of ISO's Photopolarimeter'*, Proc. SPIE **1130**, pp. 194–200.

SOFIA : STRATOSPHERIC OBSERVATORY FOR INFRARED ASTRONOMY

E. F. ERICKSON
NASA Ames Research Center, Moffett Field, CA 94035-1000, U.S.A.

(Received 24 June, 1991)

1. Introduction

SOFIA is a planned 2.5 meter telescope to be installed in a Boeing 747 aircraft and operated at altitudes from 41,000 to 46,000 feet. It will provide astronomers with routine access to infrared wavelengths unavailable from the ground, and with the means to observe astronomical objects and transient events from anywhere in the world. The concept is based on 17 years of experience with NASA's Kuiper Airborne Observatory (KAO), which SOFIA would replace. The facility would be developed and operated by NASA and the German Space Agency DARA.

2. Potential

Most infrared radiation from astronomical objects which never reaches the ground is detectable from the lower stratosphere. This fact is the principle justification for an airborne telescope. Figure 1 plots atmospheric transmission as a function of wavelength for aircraft (14 km) and mountain-top (4 km) altitudes, based on calculations by Traub and Stier (1976). The absorption is largely due to water vapour, with significant contributions from carbon dioxide and ozone in some wavelength bands. The model assumes overhead water quantities of 2.3 microns for the aircraft and 1.2 millimeters for the ground-based telescope, and a zenith angle of 60 degrees. The broad-band transmission is obviously adequate for significant photometric observations. Some of the water lines in the far infrared are still saturated at aircraft altitudes, as shown in the lower panel. Nevertheless, most important astronomical spectral features can be measured from an airborne observatory.

Features measured from the KAO are listed in Tables I, II and III. Almost all these features were first detected from NASA's airborne observatories, and cannot be studied from ground-based sites. Features seen in absorption are indicated with "a", and seen in both emission and absorption with "a/e". Whereas the dust features typically originate in regions cooler than a few hundred degrees Kelvin, the various gaseous lines originate in regions ranging in temperature from $\sim 10^5$ K down to $\sim$ 10 K. Thus these infrared spectral features characterize material in all phases of the interstellar medium: solid particles, molecules, neutral atoms, and ionized atoms.

Space Science Reviews **61:** 61–68, 1992.

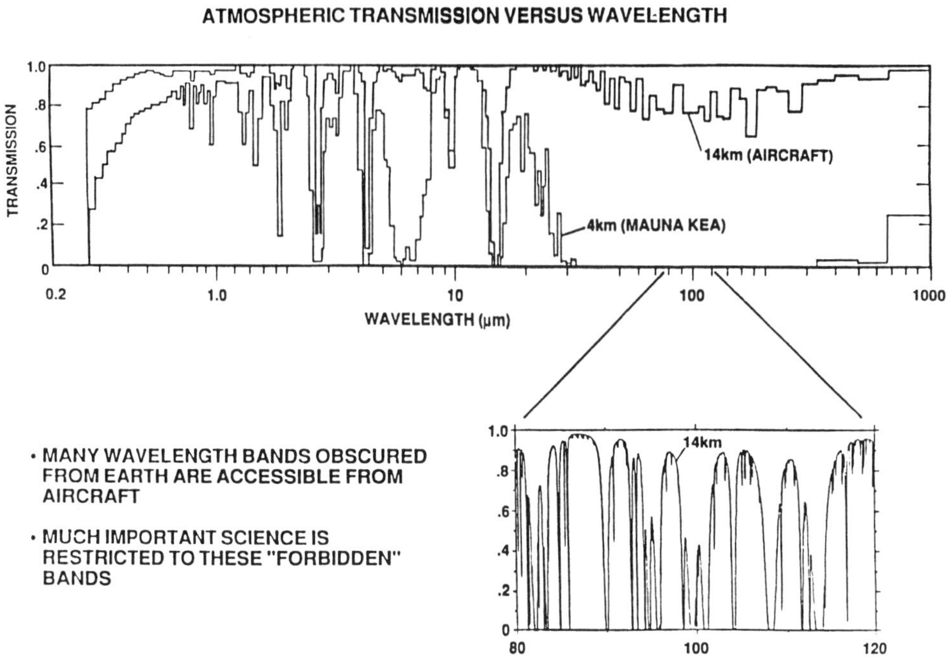

Fig. 1. Atmospheric transmission versus wavelength for aircraft and mountain top altitudes.

TABLE I
Interstellar dust grain spectral features detected from the Kuiper Airborne Observatory

Wavelength (μm)	Species
2.95?, 6.3, 7.1 a	NH_3 Ice
5.2, 5.6, 6.2, 6.9, 7.6, 7.8	PAH
5.5 a	HCO in Ice
6.0, 62, 45 a/e	H_2O Ice
6.8 a	hydrocarbon
19.0 a/e	silicate
24 - 30+	MgS (?)
45, 62 a/e	H_2O

- SOFIA -

CONCEPT: BASED ON 17 YEARS EXPERIENCE OF KAO OPERATION

CHARACTERISTICS:
- WAVELENGTH: 0.3 - 1600 μm
- APERTURE: 2.5 m
- RESOLUTION: 1 - 3 arcsec, $0.3 < \lambda < 15$ μm
  (λ /10 μm) arcsec, $\lambda > 15$ μm
- DEPLOYABLE FOR ALL SKY COVERAGE
- ROUTINE ACCESS TO INFRARED SPECTRAL REGION FOR 20 YEARS

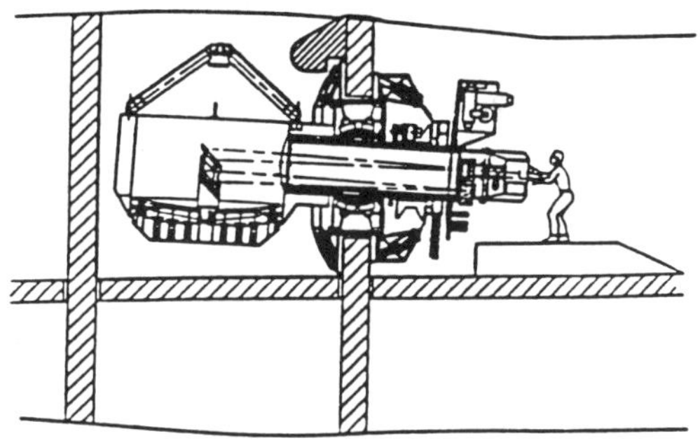

OPERATIONS:
- 120 8-HOUR FLIGHTS PER YEAR
- RAPID RESPONSE:
  ANNUAL PEER REVIEW
  EPHEMERAL EVENTS
  STATE-OF-THE-ART INSTRUMENTS
- COMMUNITY PARTICIPATION:
  ~ 15 INSTRUMENTS YEARLY
  ~ 40 P.I. TEAMS YEARLY
  HANDS-ON TRAINING FOR GRADUATE STUDENTS AND POSTDOCS

DEVELOPMENT:
- U.S.: AIRCRAFT, MODIFICATION, FACILITY, 80% OF OPERATIONS
- FEDERAL REPUBLIC OF GERMANY: TELESCOPE, 20% OF OPERATIONS

Fig. 2. Characteristics of SOFIA.

3. Facility

SOFIA's characteristics are summarized in Figure 2. The telescope must be fast and light. The design incorporates an $f/1.2$ meniscus primary with 40:1 diameter:thickness ratio, and carbon fibre reinforced plastic structure. The airbearing support and numerous other features are similar to the KAO telescope design, which permits sub-arcsecond pointing stability even in light turbulence.

Relative to the KAO, SOFIA will be roughly ten times more sensitive for compact sources, enabling observations of fainter objects and measurements at higher spectral resolution. Also, it will have three times the angular resolving power for wavelengths greater than 15 microns, permitting more detailed imaging throughout the far infrared.

Anticipated performance curves are given in Figures 3, 4, 5 and 6. On the

TABLE II
Interstellar atomic spectral features detected from the Kuiper Airborne Observatory

Wavelength (μm)	Species
1.88, 2.62, 2.67, 2.76, 4.65, 5.91, 6.6, 7.5	H I
3.1, 7.5	Ni I
4.5	Mg IV, Ar VI (?)
5.6	Mg V
6.6	Ni II
6.98	Ar II
7.63	Ne VI
8.99, 21.8	Ar III
17.9, 26.0	Fe II
18.7, 33.4	S III
22.9	Fe III
24.3	Ne V
25.2	S I
25.9	O IV
34.8	Si II
36.0	Ne III
51.8, 88.4	O III
57.3	N III
63.2, 146	O I
121.9	N II
158	C II
158	^{13}C II
370, 609	C I

sensitivity plots *PSC* and *FSC* refer to the IRAS Point Source and Faint Source Catalogues respectively.

4. Science

The infrared spectral regime encompasses a multitude of rich and varied physical processes and is uniquely suited for study of the cosmic birth on all scales. SOFIA's wavelength range from 0.3 microns to 1.6 mm includes this entire regime, and permits a very wide variety of astronomical problems to be addressed. With its high spectral and spatial resolution, SOFIA will exploit and extend the scientific legacy left by IRAS (the Infrared Astronomical Satellite). In particular, SOFIA

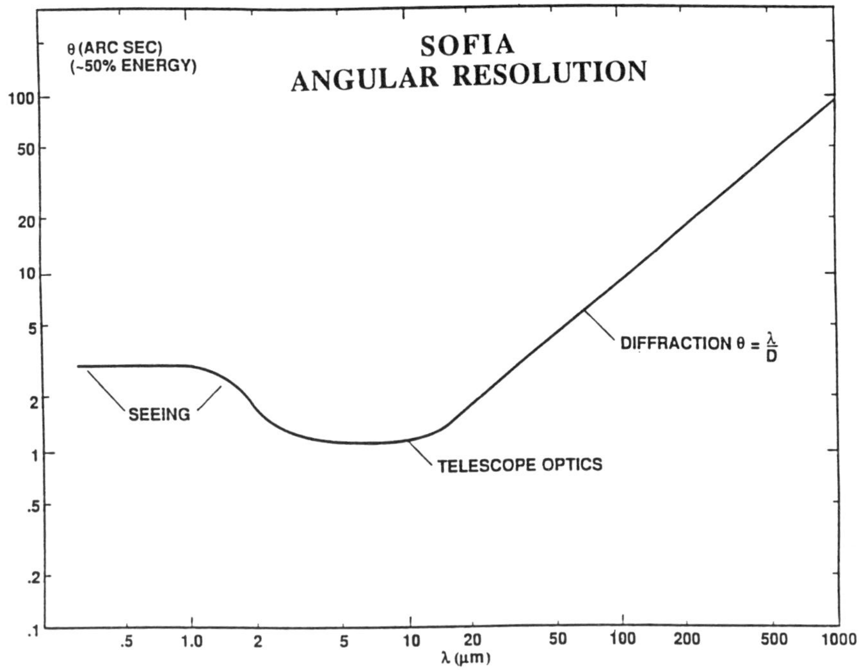

Fig. 3. The anticipated angular resolution of SOFIA.

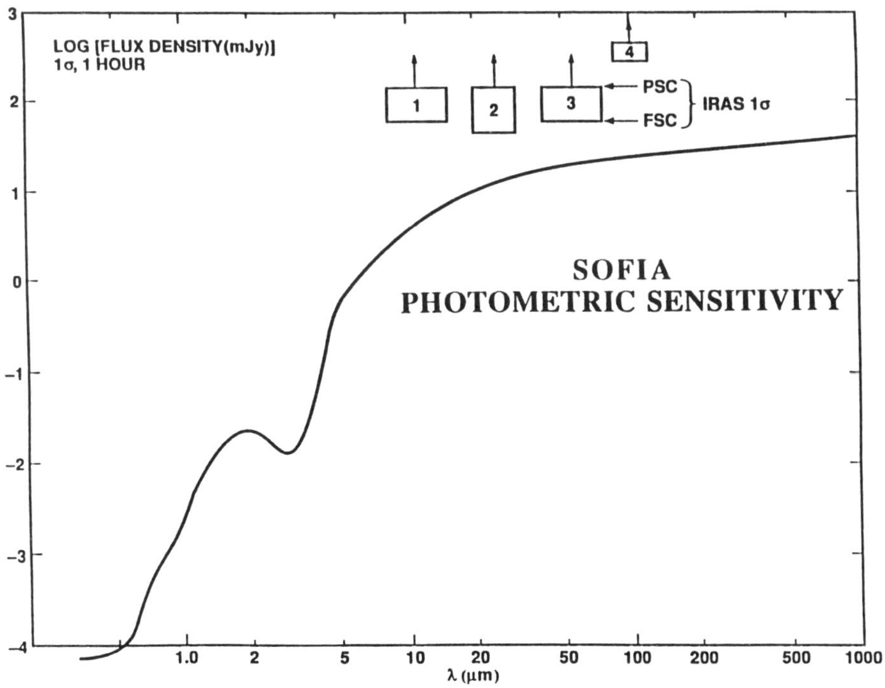

Fig. 4. The anticipated photometric resolution of SOFIA.

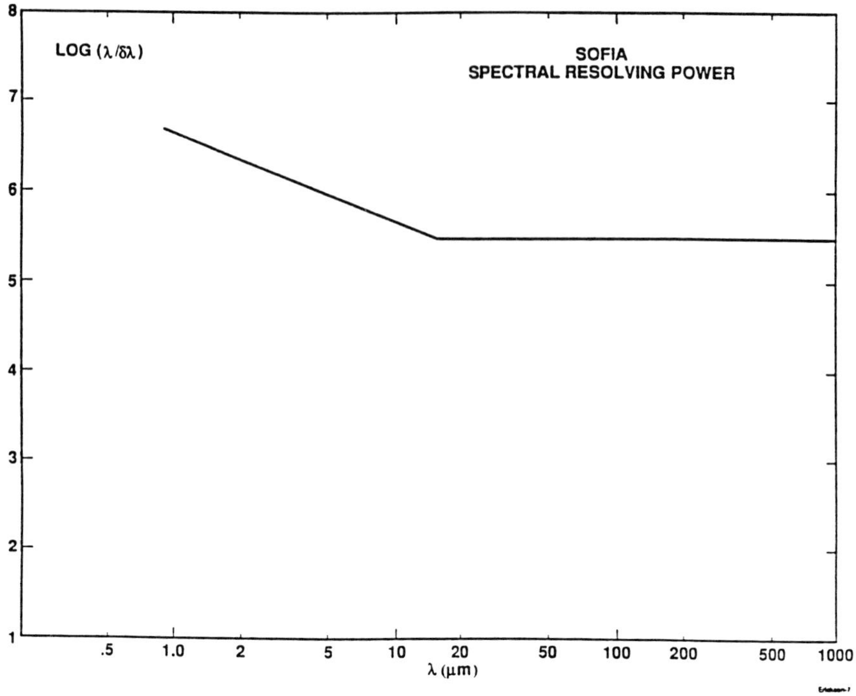

Fig. 5. The anticipated spectral resolving power of SOFIA.

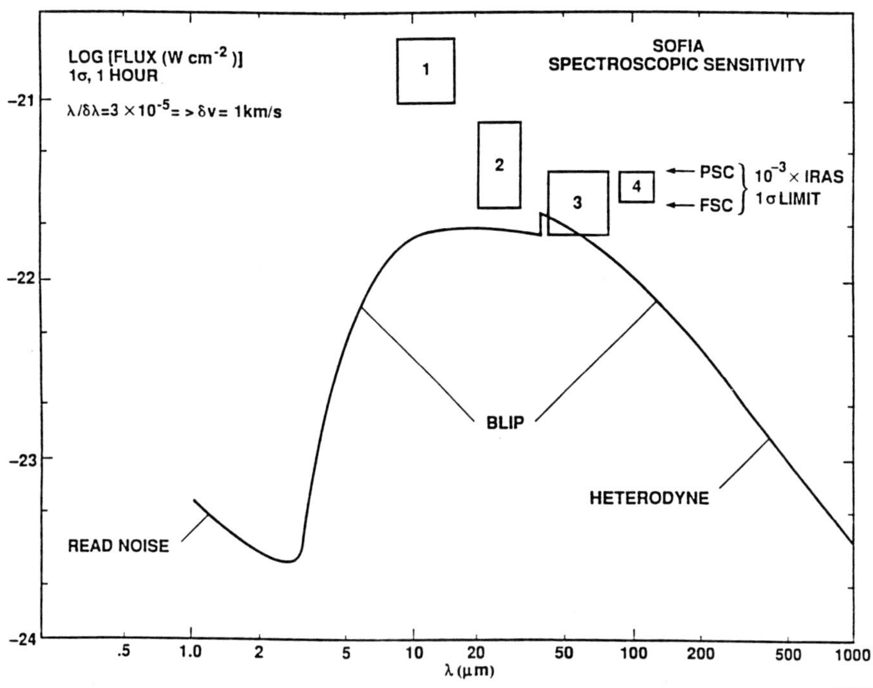

Fig. 6. The anticipated spectroscopic sensitivity of SOFIA.

TABLE III
Interstellar molecular spectral features detected from the Kuiper Airborne Observatory

Wavelength (μm)	Species
1.4, 1.8 a	C_2
2.0, 2.6, 7.5 a	C_2H_2
2.3, 4.6, 77, 84, 97, 100, 119, 124, 153, 163, 174, 186, 289, 302, 650	CO
2.9 a	C_2H
3.0, 3.5, 7.5 a	HCN
5.0 a	C_3
8.5	SiO
53.4, 84.4, 84.6, 119.3, 119.5, 163.1, 164.4 a/e	OH
308	^{13}CO
120.0, 120.2 a/e	^{18}OH
125, 166, 524	NH_3
149.1, 149.4	CH
479	HCl
790, 1600	H_2O
805	H_2D^+

will have adequate sensitivity to study any of the band 3 or 4 sources in the IRAS survey. It will complement the spectacular sensitivity for imaging and moderate resolution spectroscopy to be furnished by ISO, the Infrared Space Observatory, and by SIRTF, the Space Infrared Telescope Facility. Topics that would be addressed by astronomers using SOFIA include:

- Interstellar cloud dynamics and star formation in our galaxy: Why and how do galactic clouds form stars? How important are magnetic fields and rotation in this process?
- Proto-planetary disks and planet formation in nearby star systems: How common are solar-systems? Under what conditions are they created?
- Origin and evolution of biogenic materials in the interstellar mediums: What environments are hospitable to pre-biotic molecules and compounds?
- Comets, planet atmospheres and rings in our solar system: How did our solar-system evolve? What was the composition of the solar nebula?
- Star formation, dynamics and chemical content of other galaxies: How different are other galaxies? Why do some exhibit extremely large infrared luminosities?

– The dynamic activity in the center of our own galaxy: What powers the highly luminous phenomena hidden at the center of the Milky Way - compact star clusters? A black hole? Is this region similar to the "Active Galactic Nuclei" seen in some other galaxies?

5. Programme

SOFIA will explore these and a wide range of other science topics utilizing the unique features of its airborne predecessors: spectral coverage from the near UV to millimeter wavelengths, a routinely accessible "hands-on" working environment, annual opportunities for proposing new science and instrumentation, and deployability for all sky coverage and observations of ephemeral events such as comets, eclipses, occultations and novae.

These attributes, plus SOFIA's sensitivity, angular resolution, and 20 year lifetime, assure a vigorous and productive science programme. The broad-based community participation will include extensive opportunities for young scientists and for prompt application of new instrument technologies. At least 120, 8 hour missions per year will be flown for over 40 principal investigator teams, selected by annual peer review. Many of the investigator teams will provide the observatory with specialized instruments, including array cameras, polarimeters, and a variety of spectrometers. Thus, in addition to its wide-ranging scientific contributions, SOFIA will stimulate development of focal plane instruments, and provide important educational experience in support of future space astronomy missions.

6. Status

SOFIA definition studies, sponsored jointly by NASA and the German Science Ministry (BMFT), have been completed, and the project has been deemed ready for development by these agencies. NASA and DARA plan to begin development in 1993 if funding is available.

References

Erickson, E. F., Davidson, J. A., Caroff, L. J., and Thorley, G.: 1991, *SOFIA: A Stratospheric Observatory for Infrared Astronomy*, NASA, TM 103840

Taub, W.A. and Stier, M.T.: 1976, *Applied Optics* **15**, 364

NICMOS: NEAR INFRARED CAMERA AND MULTI-OBJECT SPECTROMETER

R. THOMPSON
University of Arizona, Steward Observatory, Tucson, AZ 85721, U. S. A.

1. Introduction

NICMOS (the Near Infrared Camera and Multi-Object Spectrometer) is a mature concept for adding near infrared capabilities to the Hubble Space Telescope that are fully corrected for spherical aberration. As shown in Section 3.1, NICMOS exceeds the infrared J and H band sensitivity and spatial resolution capabilities of both present and planned ground based facilities. The investigations discussed here utilize these unique capabilities for scientific investigations and are only possible with NICMOS.

Phase I of the NICMOS program, including the Preliminary Design Review, is complete and Phase II, normally considered Phase C and D in other NASA projects, is underway. Due to schedule and budgetary circumstances, NICMOS recently deleted spectroscopy from its capabilities. The recent 'Weedman Committee' Review, however, recommended restoration of limited spectroscopic capabilities. The form of these capabilities is under study by the NICMOS Instrument Definition Team.

2. The Scientific Programme

The NICMOS programme is a scientific investigation endorsed by two separate peer reviews. In addition to the GTO programme, NICMOS provides a versatile and powerful infrared instrument for a vigorous General Observer scientific programme. In particular NICMOS will carry out the following programmes that are not possible with any other facility.

1. Provide the deepest images of the sky in the cosmologically important J and H bands.
2. Expand our knowledge of the distance scale to the Coma cluster and beyond.
3. Provide deep, high spatial resolution, image in Paschen α which map the structure of obscured active galaxy nuclei and galactic star formation regions.
4. Provide detailed planetary imaging in infrared bands for the study of the temporal changes in planetary atmospheres.

Space Science Reviews **61:** 69–93, 1992.

2.1. Paschen α

Paschen α at 1.875 μm deserves special mention as a unique astrophysical tool. It is by far the strongest near infrared line in ionized regions. The combination of its high strength and infrared wavelength make it the most powerful tracer of the structure of ionized gas in obscured areas such as the nuclei of galaxies and star formation regions. Paschen α, however, is blocked from ground based observation by telluric water vapour. NICMOS is the only instrument capable of viewing this critical line at all, let alone with very high spatial resolution. As a single example one NICMOS Paschen α picture of NGC 1068, complementing the recent WFPC picture, will reveal the nuclear ionization regions. One more picture just off of Paschen α will reveal the nuclear stellar distribution and its relation to the WFPC discovered emission regions.

2.2. Overview

The basic scientific programme emphasizes the unique properties of HST and the near infrared spectral region. High spatial resolution observations of dust obscured regions and objects with intrinsically high infrared radiation form the basis of the programme. These include galactic nuclei, star formation regions and late stars in other galaxies. NICMOS will: more accurately determine distance scales and cosmological paraneters; produce the first high spatial resolution images of the centre of our own galaxy; and reveal the detailed structure of galactic star formation regions.

2.3. Improved Determination of the Cosmic Distance Scale and Cosmological Parameters

2.3.1. Cosmological Parameters

Many authors, (Yoshi and Takahara, 1988; Chokshi and Wright, 1988, 1989; and Cowie, Gardner, Lilly, and McLean, 1990), point out the advantages of near infrared images in the determination of the cosmological density parameter Ω_o. These advantages include the relative insensitivity to evolutionary effects, the greater galaxy flux in the near infrared, and decreased sensitivity to extinction. As an example the following two zero background simulations from Chokshi and Wright (1988) demonstrate the difference between $\Omega_o = 0$ and $\Omega_o = 1$ at $H_o = 50$. The pictures are of a $100'' \times 100''$ area with a 512×512 pixel array. This is almost exactly equal to the pixel size of Camera 3 covering an area of four Camera 3 frames. Although the pictures are for the K band NICMOS will use the H band or modified H band, just short of thermal degradation, to produce the images.

The maximum intensity in Figure 1 is 48 μJy which is 100 times brighter than the NICMOS 10 sigma limit in 1 hour of observing. The maximum intensity in Figure 2 is 130 μJy or almost 300 times brighter than the 10 sigma limit. The pictures are logarithmic covering four decades in intensity. A one hour NICMOS exposure will therefore reach all but the very faintest objects. Addition of several

FIG. 3.—Simulated image at 2.2 μm in a 100″ × 100″ field reaching a maximum redshift of $z = 10$. There are 512 × 512 pixels in all our simulated images, and the gray scale is logarithmic covering four decades in intensity. This image has $\Omega_o = 0$, $\Lambda_o = 0$, and the maximum intensity is 48 μJy pixel^{-1}.

from:Chokshi,A. and Wright, E.L. 1988, Ap.J., 333, 491

Fig. 1. Typical sky at $\Omega_o = 0$.

FIG. 4.—As in Fig. 3, with $\Omega_o = 1$, $\Lambda_o = 0$, and a maximum intensity of 130 μJy pixel^{-1}

from:Chokshi,A. and Wright, E.L. 1988, Ap.J., 333, 491

Fig. 2. Typical sky at $\Omega_o = 1$.

exposures will clearly delineate between the two cases for Ω_o.

It is very important to note that NICMOS is far more sensitive for this investigation than any other facility. A comparison with ground based observations is presented in Section 3.1. This comparison clearly shows that even an IR optimized 8 metre telescope with adaptive optics is far less sensitive than NICMOS for extended objects and that the residual seeing profile from even a 10 Zernike adaptive optic system produces confusion limited cases with far less dynamic range. The time variability of the OH background further degrades the ground based observations. In short NICMOS is the only way to carry out this programme.

2.3.2. *Cosmic Distance Scale*

The prime scientific effort of the Hubble Space Telescope is the direct measurement of the distance scale of the Universe. NICMOS near-infrared space observations have important advantages over visible and ground based infrared studies for three different distance indicators: 1) classical Cepheid variables; 2) the brightest red supergiants in a galaxy; and 3) type I supernovae. In particular, due to reduced extinction and the intrinsically higher infrared brightness of these indicators, the near-infrared is more precisely calibratable and more sensitive than the visible spectral region. Also, NICMOS is intrinsically more sensitive to point objects than ground based facilities. The lack of a residual 'seeing' component in the point spread function (see Section 3.1) is essential for these studies.

NICMOS links the well-established distance indicators such as the Cepheids in nearby galaxies (Local Group) to the brightest red supergiants, (readily observable in the Virgo cluster), and to Type Ia supernovae . NICMOS easily reaches the brightest red supergiants in pure Hubble flow galaxies at the distance of the Coma Cluster, which significantly overlap the observable range of Type Ia supernovae (out to galaxies at a $z = 0.25$). In this way NICMOS both validates and calibrates secondary distance indicators such as supernovae for use at the greatest distances.

The key to this programme is the calibration of the brightest red supergiants by observing Cepheid variables of known type and the brightest red supergiants in the same galaxy. NICMOS extends the expected WFPC II study of Virgo cluster Cepheids to regions of higher extinction. These regions are also the regions of highest red supergiant concentration. The ratio of visible (V) to infrared (K) extinction is 11 in magnitudes. Thus a Cepheid with 11 magnitudes of visible extinction has only 1 magnitude of extinction at 2.2 μm. NICMOS is also faster than the WFPC II for long period variables detecting a 20 day Virgo Cepheid ($M_V = -5.0$) at a S/N of 10 in 1000 seconds. Virgo Cepheids play a key role in the Cepheid–brightest red supergiant distance indicator link.

NICMOS possesses even more pronounced advantages for observations of the brightest red supergiants. The visual luminosity of the brightest M supergiants is near $M_V = -8$ while their luminosities at H and K are near -12 mag. The red supergiants are intrinsically much brighter in the infrared and the uncertainties of interstellar extinction are greatly reduced. The infrared luminosity is also directly correlated with the bolometric luminosity. The brightest M supergiants populate the observed upper luminosity boundary for evolved supergiants near $M_{bol} = -9.5$ mag. This is attributed to a stability limit which prevents the evolution of massive stars ($> 50 M_\odot$) to cooler temperatures.

The brightest red supergiants are reliably distinguished from foreground M dwarfs by their J-H and H-K colors. Narrow band CO photometry further enhances their distinguishability. The absolute bolometric luminosity (and M_H and M_K)

of the brightest red supergiants shows only a slight dependence on the intrinsic luminosity of the parent galaxy. However, it is not distance degenerate, and for spirals and Magellanic irregulars ($M_V < -19$) the calibration of M_{bol} (and M_H and M_K) has a dispersion of less than 0.2 mag. We will recalibrate the modest dependence of absolute bolometric luminosity of the red supergiants by 2 μm observations of Cepheids and red supergiants in the same galaxies at Virgo-like distances. The most reliable determinations will be based on an ensemble of the brightest red supergiants rather than the single brightest.

2.4. Extragalactic Astronomy

The programme of extragalactic observations is of prime interest to the NICMOS team. The following discusses some aspects of that programme.

2.4.1. High Luminosity Galaxies and Hidden QSOs

Many high luminosity infrared galaxies have bolometric outputs that rival QSOs. The luminosities of these galaxies exceed 10^{12} $L_\odot$, and their number densities exceed those of QSOs in the same low redshift volumes. These two facts suggest that these galaxies harbour dust-enshrouded QSOs in their nuclei which evolve into more traditional QSOs as their dust shrouds are blown away. The specific mechanism for creating QSO nuclei is hinted at by the observational evidence that virtually all of these galaxies have indications of interactions or mergers.

A QSO formation model which fits these data is a nuclear starburst induced by the interaction of two gas-rich spirals. The starburst produces a centrally condensed stellar cluster and gravitationally trapped gas feeds a central black hole which becomes the QSO. The study of low redshift, high luminosity galaxies provides important information on star and AGN formation processes at higher redshifts.

NICMOS high spatial resolution observation of low redshift high luminosity systems are especially meaningful for checking theories which predict mergers and interactions between galaxies in the early universe. Since the evidence for these mergers is contained in the dust shrouded nuclei of these galaxies NICMOS is essential for these observations.

2.4.2. Extragalactic Star Formation

A unique capability afforded by NICMOS is the measurement of molecular and atomic hydrogen distributions at high angular resolution in moderately or heavily obscured extragalactic star formation regions. Gentle shocks produce strong emission in H_2 lines such as 1 – 0 S(1) (2.122 μm), 1 – 0 Q(1) (2.407 μm), and 1 – 0 Q(3) (2.424 μm) and H II regions emit strongly in Paschen α at 1.875 μm. It has been suggested that supernova-driven shock phenomena play a central role in the propagation and eventual extinction of starbursts. The resolution of NICMOS determines the shock structure on the scale of moderately old supernova remnants (e.g., a shell expanding at 1000 km/sec at a distance of 3 Mpc reaches a diameter of 0.2$''$ in 1500 years).

2.4.3. Supernovae

Supernova 1987A in the Large Magellanic Cloud revolutionized the study of supernovae in general and will continue to be an observable HST target for many years. For example, it is estimated that the power supplied by radioactivity in the interior will remain above 1036 ergs s^{-1} (250 solar luminosities) for 6 years and a possible pulsar may maintain 100 times more power for at least 10 years. Interactions of the expanding remnant with the surrounding environment will remain observable on much longer time scales.

Scientific issues to be addressed through imaging of SN 1987A include: (1) investigation of the dynamical evolution of the expanding remnant; (2) possible probing of the degree of inhomogeneity of the circumstellar matter ejected through the previous history of the precursor star; and (3) the study of the chemical composition and evolution of the supernova as it develops into a unique cold cloud ($T \leq 100$ K) that harbours a radioactive energy source and possibly also a compact X-ray source. In this last instance, infrared narrow band photometry of the most abundant molecular species (e.g. CO and H_2) is particularly important. Moreover, should the supernova form dust, infrared studies with high angular resolution will be invaluable.

2.4.4. The Galactic Centre

The Galactic Centre has been studied from the ground extensively using newly developed arrays. The very crowded central parsec is surrounded by an extensive region of elevated star counts whose edge is seen only in very large scale image mosaics. Whether the bright stars seen all over this region merely represent the tip of the asymptotic giant branch or represent a younger population remains to be determined. Many of the issues in the central parsec such as the size or very existence of a black hole as well as the nature of the stars are still unsettled after much effort from the ground. Resolution of the puzzles presented by the Galactic Centre requires the high spatial resolution of the NICMOS cameras.

2.4.4.1. High Resolution Imaging. A careful measurement of the surface brightness distribution as a function of radial distance from the Galactic Centre constrains the size of a black hole located at the centre. If a black hole is present at the Galactic Centre, it should lie somewhere near Sgr A. Extreme crowding in the Galactic Centre region requires the NICMOS imaging capability to eliminate source confusion especially in the IRS 16 complex. The objects comprising IRS 16 are not easy to model. Spectral evidence indicates that they are not late-type stars like many of the other sources, and photometric evidence indicates that they are nearly the bluest sources at the Galactic Centre. The high velocity gas winds are centred on this complex.

A more convincing possibility is that these sources are actually small groups or clusters of O stars, similar to 30 Doradus. O star clusters fit the colour and luminosity data and also provide much of the ionizing flux present. The cluster model is able to explain the stability of stellar groups in the Galactic Centre environment whereas a black hole lying only 0.1 pc from the groups would tend to disrupt them. NICMOS images with 0.075 arc second pixels will check the cluster hypothesis and also measure the nuclear luminosity profile of unresolved stars.

2.4.4.2. Astrometry of Galactic Centre Stars. A major uncertainty in deriving the mass distribution at the Galactic Centre from radial velocity measurements is the lack of information on the shapes of the stellar orbits or equivalently, the stars' true space velocities. Determination of the proper motions of the fastest moving stars will give the true space velocity and thereby reduce the modelling uncertainties in the mass distribution. The stellar velocity dispersion is 80 km/sec so if 100 km/sec is taken as a representative velocity, a star's apparent position at the 8 kpc distance of the Galactic Centre would change by $.0026''$ in one year or by $.013''$ in five years. Averaging results from several exposures will permit proper motions to be derived from data taken only a year apart. Smaller velocities and/or higher accuracies are derived from data taken over longer time periods. Note that there are so many sources in the Galactic Centre that a local reference frame can be set up for these measurements without dependence on astrometry from the Fine Guidance System.

2.4.5. High Spatial Resolution Infrared Observations of Young Stellar Objects
The NICMOS programme concentrates on observations of the physical processes which control the evolution of Young Stellar Objects (YSOs) from the time they first emit significant near-infrared radiation to the time they become easily observable optical objects. The large camera fields make surveys of selected star forming regions both feasible and very attractive.

2.4.5.1. Deep Surveys of Star Forming Clusters. The goal of this programme is reliable statistics for the study of low mass star formation in representative bound clusters and unbound associations. Complete luminosity functions in the core regions of four different, nearby star forming regions provide a comparative study of the star formation process down to the lowest masses. For this investigation, the superlative spatial resolution and high point source sensitivity of HST are essential. Initial surveys from the ground with infrared array cameras are troubled by unresolved multiple sources in many fields. The HST angular resolution minimizes problems from source crowding and easily identifies close multiple systems. To extend our knowledge to the lowest masses, we require the sensitivity and areal coverage of the NICMOS cameras.

The basic data base for the programme is a complete survey of selected regions of the Taurus, Ophiuchus, Orion, and NGC 2264 star forming regions. The surveys will be complete to K, H, and J magnitudes of +20 in fields of four square arc minutes in Ophiuchus and Taurus and two square arc minutes in Orion and NGC 2264. These infrared surveys are the best way to obtain a census of the embedded stellar population in these obscured regions. Even in the presence of 55 magni-

tudes of visual extinction, the survey finds all stars of spectral type M8 or earlier. Extrapolation of current source counts predicts over 800 objects in each survey region. From these large samples, it is possible to construct reliable, very complete luminosity functions.

2.4.5.2. Multiple Star Systems. This programme determines the effect of mass on pre-main sequence evolution by measuring the properties of individual objects in tightly bound multiple systems. Very close spatial association makes a common age and chemical composition for the objects in the group highly likely. By studying the differences between these objects, we obtain a better understanding of the role of the remaining variable, mass. The survey photometry produces evolutionary tracks for these pre-main sequence stars.

This programme uses the same basic data as the deep survey described in the previous section. From the deep surveys of star forming regions, the most promising close multiple systems are selected. The methodology involves photometric follow-up with a statistically significant number of observations and narrow band images in H_2 and Paschen α.

2.4.5.3. Morphology of YSOs. Ground-based observations of YSOs show a rich amount of structure. Bipolar CO molecular flows extending out to 1 pc are characteristic of the very youngest stellar objects, while high density tracers such as CS and NH_3 show flattened structures on the tenths of pc scale. Observations in optical emission lines reveal highly collimated jets on much smaller scales of 100 - 10000 AU. Finally, infrared observations suggest the presence of flattened, disk-like structures which are interpreted as scattered light from dust disks. The advent of infrared array cameras has made possible much higher quality studies of these infrared structures.

It is now clear that the outflow phenomenon is an important and perhaps necessary part of the star formation process. Still unclear, however, are the exact nature of the collimation mechanism, and the interrelationship between the various structures. The high angular resolution of HST images is essential for resolving structures over a critical range of sizes to test competing outflow models.

Polarization images are particularly valuable in understanding the detailed structure of the sources. The position angle is typically perpendicular to the molecular outflow axis. These observations are consistent with the bulk of the near infrared radiation emerging as scattered light from a collimating disk. Alternatively, the radiation in L1551 IRS 5 can be due to an unresolved biconical reflection nebula. Although the most recent observations favour the model where the infrared radiation scatters off the disk, the definitive observations require the improved angular resolution possible with HST.

2.4.5.4. Interaction with the Surrounding Medium. Inflow and outflow, disks, jets, shocks, H II regions, molecular hydrogen or CO emission are all examples of

interactions which are important in the evolution of a YSO to the main sequence. Recently these mechanisms have become the subject of intense observational and theoretical study. Observations are showing that the hydrogen line excess problem extends to even low mass YSOs, that infrared CO emission is far more ubiquitous than previously understood, and that indirect evidence is mounting for disk structures being very common. New theoretical programmes relating the infrared line emission to the basic properties of YSOs have been initiated. The structures of interest in the following programme lie within a 1 – 2 arc second radius of the central object.

High spatial resolution observations of the strong Paschen α line at 1.875 microns are very important new tools in this investigation. In the Taurus and Ophiuchus regions at 150 pc, the spatial resolution is 10 AU. Direct imaging of the YSO's associated with molecular outflow sources tests various theories for the collimation of the flows. In one class of theory, an initially spherical stellar wind is focused by density perturbations in the surrounding medium. In this case, the wind clears out a region in the vicinity of the YSO. In another class of theory, the outflow comes directly from the disk and derives energy from the accretion process. In these models, there is a transition between ionized and neutral flow on a size scale of $\sim 10^{15}$ cm ($0.5''$ at Taurus).

High angular resolution NICMOS images will show the relationship between the disk, the large molecular outflow, and the highly collimated flows indicated by optical emission lines. Similar imaging in infrared emission lines investigates the excitation structure of these jets with minimal complication due to the dust extinction that is always present in these sources.

The freedom from atmospheric absorption and airglow emission, coupled with the high angular resolution of NICMOS, are special advantages for the study of H_2 emission from shocks in star-forming regions. H_2 is expected to be a dominant coolant of shock heated molecular gas and thus serves as a primary probe of shock structure. Very high angular resolution is needed to reveal detailed shock structure. Freedom from atmospheric effects also makes NICMOS better suited for detecting extended, low surface brightness emission. A project of major importance is the measurement of an interstellar shock with sufficient angular resolution to infer the variation of conditions with projected distance from the shock front. These observations make direct tests of theory concerning the structure of interstellar shocks in ways that have not been previously possible. The cooling length of a 12 km s^{-1} shock propagating through a neutral cloud of preshock density n_0 is $l_c \sim 2.8\ 10^{17}/n_0$ cm for the cooling from T = 5000 to T = 4000 K. For a shock front viewed edge-on at a distance of 1 kpc, this corresponds to an angular extent of 0.2 arcsec for $n_0 = 100$ cm^{-3}.

2.4.5.5. Infrared Cirrus and Circumstellar Dust. The infrared cirrus consists of mixtures of interstellar dust grains and large molecules, PAHs, heated by the ISRF. Under typical interstellar conditions, thermal reradiation of absorbed starlight oc-

curs at temperatures from 20 K to several hundred K and the particles scatter the ISRF across the UV, optical and near-IR spectrum. These grains and complex molecules form in ejecta of late type stars where they can first be detected by scattered starlight or by thermal reradiation. By combining observations of both mechanisms over all wavelengths of interest for various interstellar and circumstellar conditions we can greatly expand our knowledge of the solid materials from which stars and planets are formed. Comparisons of NICMOS multi-band images of close circumstellar material and infrared cirrus regions are keys to this study.

2.4.6. Brown Dwarfs in the Solar Neighbourhood and in Young Star Clusters

The wide field, infrared detectors and coronagraphic capability of NICMOS combined with the imaging characteristics of HST permit sensitive searches for relatively cool, compact objects in orbit around nearby stars and stars within young clusters. This material is likely to include red dwarf stars near the hydrogen-burning mass limit (0.085 $M_\odot$) and substellar brown dwarfs (0.010 – 0.080 $M_\odot$). Photometrically calibrated images of such objects are essential for calibrating the lower main sequence, discriminating between various evolutionary models of low mass objects, and determining the role of brown dwarfs in planetary systems. The NICMOS cameras and coronagraph accomplish high spatial resolution imaging with much greater dynamic range than can be achieved from the ground. This performance is the direct result of low noise detectors, an optical design which achieves a stable modulation transfer function, and the elimination of variable atmospheric seeing as well as background radiation from airglow, thermal emission, and scattering.

The hunting grounds for brown dwarfs derive from several strategies and questions to be answered. The first strategy includes the most likely areas to find brown dwarfs, the second is directed toward determining how many brown dwarfs exist. The following are several likely areas of search.

ASTROMETRIC 'BINARIES' – Astrometric binaries are useful as hunting grounds for brown dwarfs, as objects for calibrating the lower main sequence and for testing theoretical models of stellar evolution. The photocentric orbit from astrometry plus direct imaging at a single epoch yields the absolute magnitudes and masses of the individual components and also provides valuable age information on the system.

YOUNG STAR CLUSTERS – The luminosity of a brown dwarf is primarily from thermal radiation of heat generated from gravitational contraction and from a burst (10 Myr) of deuterium burning (for objects $> 0.015\ M_\odot$). Therefore, young objects can considerably outshine their older counterparts, and it is reasonable to search for them in young star clusters such as the Hyades (45 pc; 0.7 – 0.8 Gyr) and Pleiades (120 pc; 0.007 Gyr). For example, a 20 Jupiter mass object in the Pleiades is expected to be as luminous as a 40 Jupiter mass (5 Gyr) object at a distance of 5 pc.

RADIAL VELOCITY VARIABLES – Several observatories are searching for extra-solar planets around solar type stars by detecting stellar radial velocity per-

turbations. These measurements are capable of sensing the presence of orbiting Jupiter-like planets. In fact, several solar-type dwarf stars exhibit such perturbations after only 5 years of data. However, it is not possible to determine the component masses independently of the effects of orbital inclination. Direct imaging can exclude the possibility of a more massive companion at high inclination and can discover companions missed in face-on orbits. In systems with positive detections from both methods, the orbital elements can be determined more accurately leading to accurate mass information.

To determine the density of brown dwarfs it is reasonable to search nearby stars.

SURVEY OF NEARBY STARS – A high angular resolution survey of nearby stars within 100 pc offers several advantages over previous and on-going surveys at lower resolution. Ground-based imaging at 1 – 2 arcsec resolution yields a relatively coarse (100 – 200 AU) spatial resolution at 100 pc. It is clearly important to examine the environs of stars on solar system sized scales since accretion processes may play an important role in the formation of substellar objects, as indicated already by the existence of the companion to HD 114762. Considering other, indirect search methods (astrometry and radial velocity perturbations), the effects of very long revolution periods and orbital inclination can preclude detection of a companion. In addition, the astrometric method favours detection of substellar objects around red dwarf stars. That many companions may be missed by these methods is demonstrated by the detection of Gl 866B and by observational results on astrometric binaries showing no separations $>$ 1 arcsec.

2.4.7. Beta Pictoris Systems

The Infrared Astronomy Satellite observed 25 A0 – K5 stars with substantial ($> 3\sigma$; $T > 35K$) infrared excesses in bands 2, 3, and 4. These stars are surrounded by gas and dust – protoplanetary material which possibly accreted planetary systems near the central star. The dust particles are known to be large (0.8 microns to 1.1 mm), neutrally scattering, and apparently in orbit around the central star. Many main sequence stars have such particle clouds but their existence is not strongly correlated with properties of the star such as age, spectral type, binarism, or rotation. The circumstellar disk surrounding the nearby (16 pc) star, Beta Pictoris (sp AV, K = 3.5 mag) is the most easily observable of all such objects known at this time. Because the Beta Pictoris disk may represent an early stage in the development of our own solar system and, furthermore, appears to be a massive analogue to our inner Oort cloud, it holds a special interest with regard to its physical and compositional properties.

Two key NICMOS areas of study for Beta Pictoris type systems are the size distribution and composition of the disk material. Scattering properties change rapidly at wavelengths equal to the size of the scatterer. The near infrared wavelengths are equal to the sizes of the smallest particles. Narrow band imaging will also determine the material composition via the broad band solid absorptions in the NICMOS spectral region.

2.4.8. *Solar System Astronomy*

The following areas of solar system study deserve particular mention. In addition to the GTO program we expect NICMOS to become a prime instrument for GO observations by the planetary community.

2.4.8.1. Atmospheres of the Outer Planets. High infrared spatial resolution gives NICMOS a unique capability to study physical and chemical processes at work in the giant planets. Two types of observations are especially important. First, spatially resolved observations of cloud features yield direct tracers of atmospheric dynamics. These observations examine and test models of seasonal and latitudinal variations in the absorption of solar radiation, modes of energy transport of internal heat sources, and the distribution of rotational angular momentum with depth. Second, vertical profiles of major atmospheric constituents help constrain molecular and elemental composition providing key inputs for theoretical models of the formation of the giant planets and the solar system. These profiles also study the different cloud layers which play a primary role in the absorption of solar radiation and also the transport of thermal energy from the deep interior. Such spatially resolved observations are not possible with ground-based instruments.

With regard to the first class of observations, dynamical models of the atmospheres of Jupiter, Saturn, Uranus, and Neptune are generally based on centre-to-limb variations (CTL) within specific spectral bands. It is essential to isolate individual atmospheric features in CTL scans in order to avoid averaging over diverse structures. NICMOS yields resolutions of < 700, 1400, 3000, and 4800 km for these planets, respectively. By monitoring such features in molecular absorption bands of CH_4, NH_3, and hydrogen as they move from centre to limb, NICMOS samples a large range of depth in the atmospheres which provides information not available from HST first generation instruments or from the Voyager cameras.

For the second type of observations, the narrow band photometry capabilities of NICMOS will be used to observe in atmospheric bands of methane, ammonia and hydrogen. These bands will be free of telluric absorption features, allowing direct comparison with laboratory observations at relevant temperatures and pressures. Detailed comparison with laboratory data elucidates abundances and haze structures (which affect the radiative transfer in the lines). For Jupiter and Saturn, the methane observations determine the structure and particle properties of the tropospheric haze (i.e. ammonia clouds) with good spatial resolution through variations in the depths of methane absorption features with different intrinsic strengths. Similar information is derived from the hydrogen bands. The ammonia absorption bands determine the vertical distribution of ammonia over a range of altitudes.

For Uranus and Neptune, high quality methane images with no telluric interference improve the determination of the vertical profile of methane and the properties of methane clouds. Spatial, diurnal, and seasonal variations in the properties of these hazes will be assessed with repeated observations over time. Such observations are especially important for Neptune which has exhibited bright and variable cloud

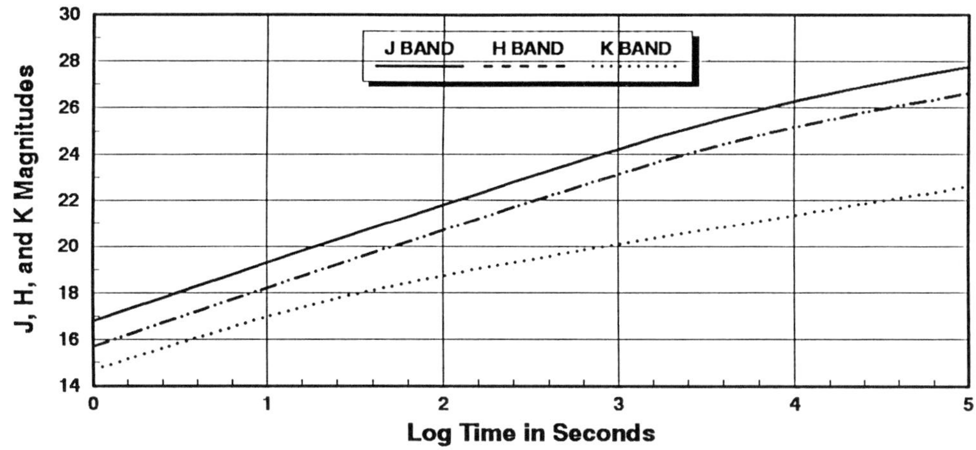

Fig. 3. The performance of NICMOS Camera 1.

structures as well as substantial changes in the distribution of reflecting material in its northern hemisphere.

2.4.8.2. Surface Compositions of Small Bodies. High spatial resolution NICMOS narrow band imaging will help in understanding the complexity of the surface materials on Neptune's Triton. NICMOS cameras enable further progress in elucidating the physical states of nitrogen and methane on its surface (as well as search for the cosmochemically important carbon monoxide). These tasks still remain after the 1989 Voyager flyby since that spacecraft, with the exception of a broadband-visible methane filter, did not have instruments capable of diagnosing the surface composition of frosts. As the subsolar point on Triton moves rapidly to high latitudes over the next ten years, periodic observations by NICMOS will allow inference of the seasonal transport of methane and other volatiles across Triton's surface.

3. Performance

NICMOS performance levels are set by the natural, telescope, and instrument backgrounds, the instrument efficiency, and the detector performance. With the HST background levels shown in Figure 6 and the parameters listed in Table II Figures 3 – 5 demonstrate the NICMOS camera performance levels.

3.1. Comparison with Ground Based Facilities

The spatial resolution and source sensitivity of NICMOS greatly exceeds any current ground based facility. However, since the inception of the NICMOS programme there has been significant progress in studies of adaptive optics and their

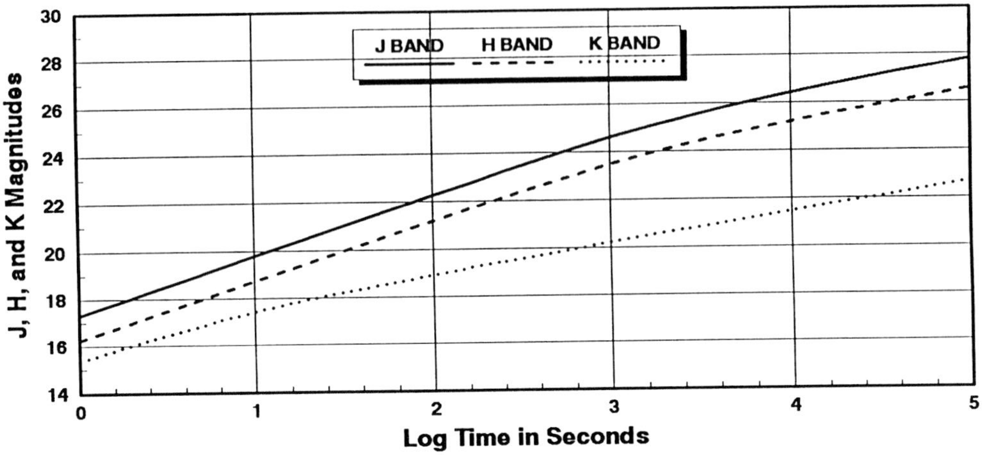

Fig. 4. The performance of NICMOS Camera 2.

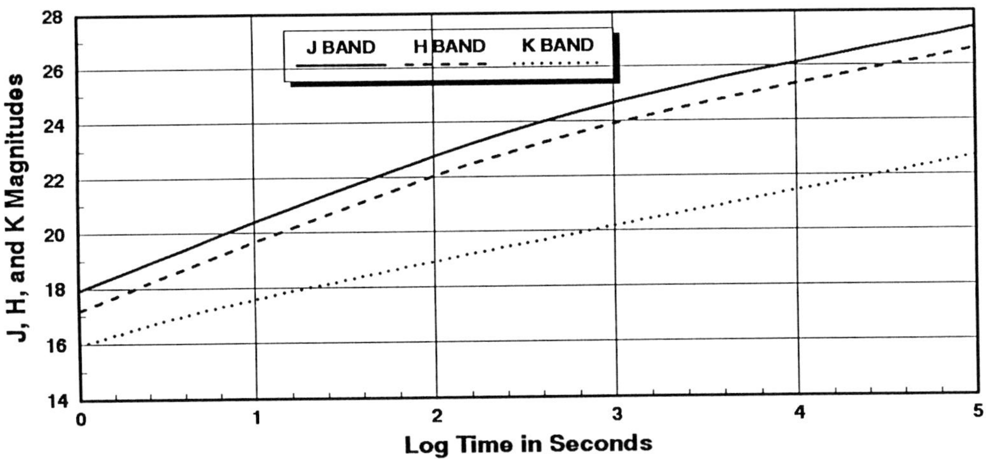

Fig. 5. The performance of NICMOS Camera 3.

possible application to future large ground based telescopes. Even so, several factors prevent ground based telescopes from matching the J and H band sensitivity and spatial resolution of NICMOS. These factors, discussed below are:

1. Bright OH sky glow for ground based observations;
2. Fast temporal variations in the OH sky glow;
3. Residual seeing components in a 10 Zernike adaptive optic correction;
4. Very fast temporal variations in the seeing residual;

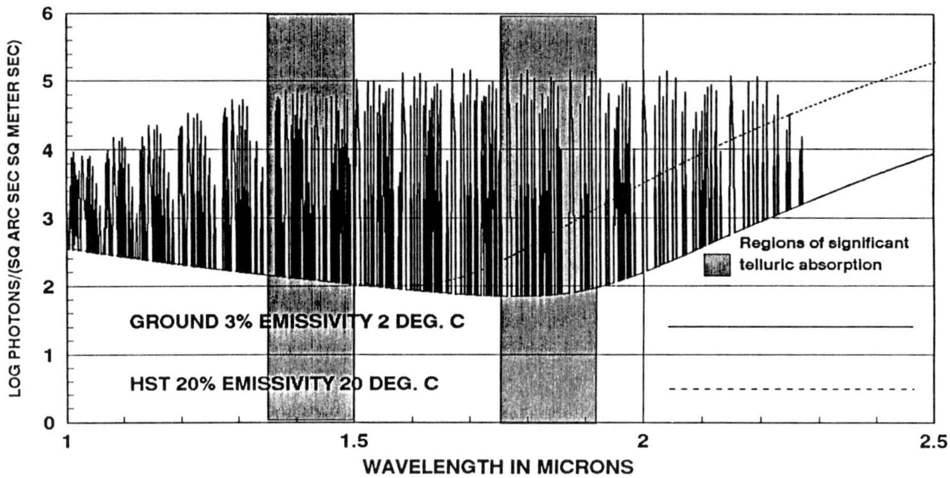

Fig. 6. Telescope and sky background spectrum.

5. The very low percentage of time that a full 10 Zernike correction can be achieved.

An IR optimized, 8 metre ground based telescope with adaptive optics, similar to that considered by various advisory panels, serves as a basis for comparison. We assume that adaptive optics capable of a 10 Zernike term correction are operating, although Table IV indicates that this can rarely be achieved even at the best of sites. In addition stars of sufficient brightness to provide the reference point spread function may be only available over a small percentage of the sky.

3.1.1. Sky Background

The first factor for consideration is the sky background in the 1.0 to 2.5 μm range. Figure 6 shows the background in (photons/sec)/(square metre and square arc second) for HST with 20% emissivity at 20°C and a ground based telescope with 3% emissivity at 2°C, (the average Mauna Kea temperature). Although these are commonly used emissivities we believe the ground based number is as optimistic as the HST number is pessimistic. The primary ground based background out to 2.2 μm is the myriad of OH emission lines which are completely absent for NICMOS. The other background components are the scattered zodiacal background at short wavelengths and the telescope thermal emission at longer wavelengths.

Table I gives the integrated background in the J, H, and K photometric bands. The ground 20% emissivity column represents current ground based telescopes.

TABLE I
Sky backgrounds in mag/sq arc second

	HST	GROUND BASED	
	20% emissive	20% emissive	2% emissive
J	20.07	16.33	16.33
H	19.45	14.80	14.80
K	12.18	13.18	14.15

3.1.2. Sky Variability

The following analysis assumes that the background noise in the ground based case is simply the square root of the number of detected background photons. However, it is well known that the OH sky emission is variable on time scales of a few minutes or less. If the sky background is removed by 'background' observation at a different time, it may be possible to only reduce the background noise to a few percent of the ground based background. This is a severe disadvantage for ground based observations.

3.1.3. Point Source Sensitivity

The source of the NICMOS advantage in the J and H bands is obvious from Table I. To show quantitatively this advantage Figures 7 through 9 compare the performance of NICMOS, current ground based telescopes and an IR optimized 8 metre telescope. The 8 metre performance has been set to the performance levels derived by the various advisory panels considering it. A summary of the telescope parameters follows in Table II.

The total angular size is the linear size in arc seconds of a square area of integration. All of the background incident on the area is included and only the source flux from the telescope point spread function is included. The usual FWHM seeing angular size is roughly half of the total angular size. For example, the 0.2 arc second case is roughly equal to a 0.1 arc second FWHM. The telescope point spread function referred to is that appropriate for the telescope diameter and central obscuration. In the case of NICMOS not all of the source flux falls in the integration area. Justification for the angular sizes is given in the next section.

The performance graphs indicate the point source magnitude that gives a signal-

TABLE II
Comparison of NICMOS and ground based telescopes.

	NICMOS	8 Metre Ground	3.6 Metre Ground
Effective area	4.07 sq m	47.75 sq m	9.92 sq m
Optics temperature	20°C	2°C	2°C
Emissivity	20%	3%	20%
Total efficiency	30%	30%	30%
Detector noise	25 electrons	25 electrons	25 electrons
Dark current	1 electron/sec	1 electron/sec	1 electron/sec
Total angular size	0.2 arc sec	0.2 arc sec	1 arc sec

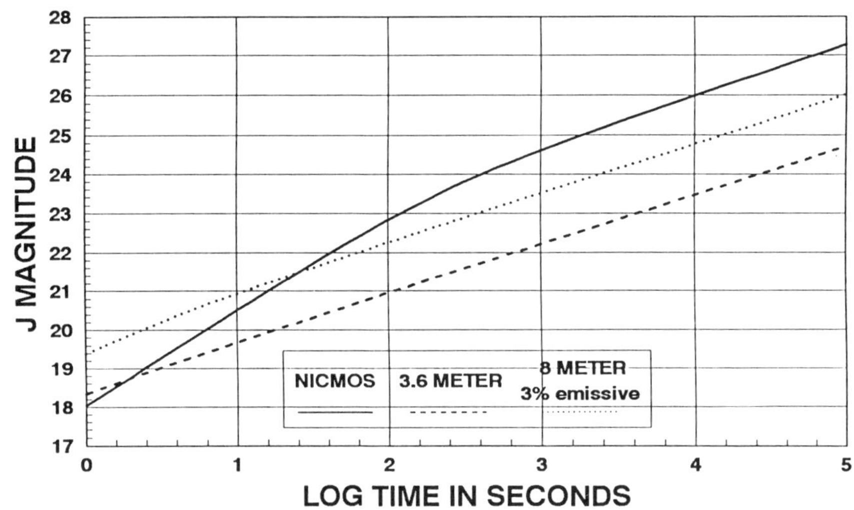

Fig. 7. Relative performance in the J band.

to-noise ratio of 10 after the indicated integration time for each of the three cases. These graphs show that NICMOS exceeds the J and H sensitivity of both current telescopes and a future 8 metre IR optimized telescope with adaptive optics but is not competitive at K.

3.1.4. Extended Source Capability

Many of the important NICMOS observations are of extended objects. Table III displays the relative performance for the detection of an object subtending 1 arc second under the parameters given in Table II.

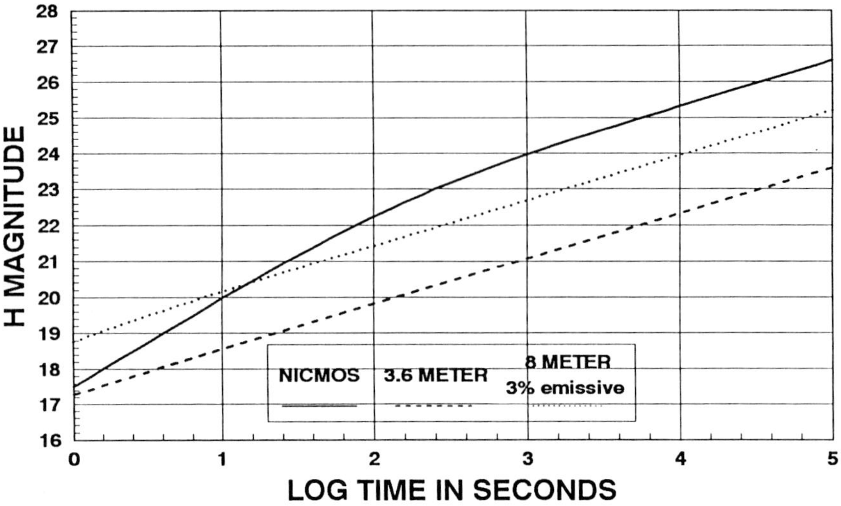

Fig. 8. Relative performance in the H band.

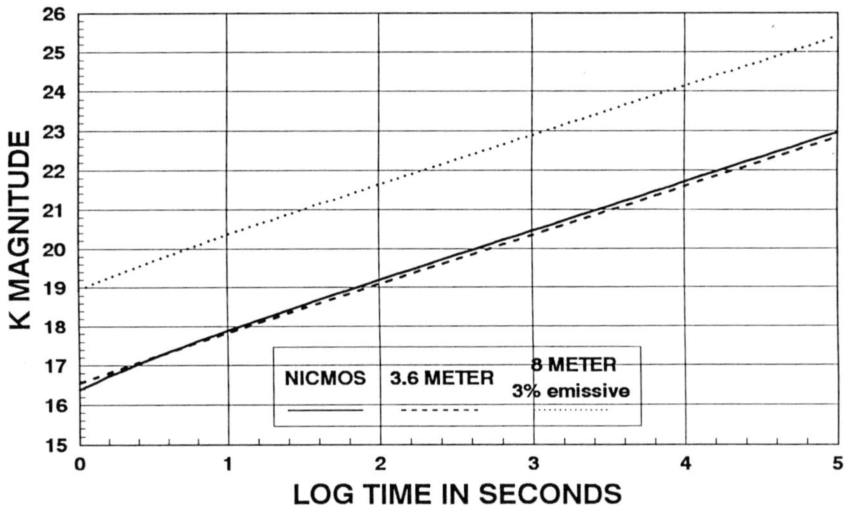

Fig. 9. Relative performance in the K band.

Again the NICMOS sensitivity exceeds the ground based numbers in the J and H bands. The ratios are higher in this case since the smaller pixel NICMOS case did not include all of the flux and was not background limited. NICMOS can still benefit from improvements in dark current and read noise, whereas, the ground based measurements are completely background limited. The NICMOS observation is the sum of 25 0.2 arc second pixels. The ground based number assumes a single 1 square arc second pixel, but, it is not relevant since the observation

TABLE III
Sensitivity to extended sources

LIMITING MAGNITUDE FOR ONE SQUARE ARC SECOND

	J	H	K	PIXEL SIZE
NICMOS	23.09	22.94	19.43	0.2
8 metre	21.64	21.06	21.31	NOT REL.

signal to noise = 10.
integration time = 1000 seconds.

is completely background dominated. Again the ground based numbers assume a constant background. OH variability will greatly degrade the ground based facility from that of Table III.

3.1.5. Angular Resolution

A critical comparison between NICMOS and large ground based telescopes is angular resolution. Although the intrinsic diffraction limited point spread function for an 8 metre telescope is narrower than that of the 2.4 metre HST, the diffraction limited point spread function can not be achieved from the ground even with adaptive optics. Beckers (1988) describes the point spread function with adaptive optics as the sum of a diffraction limited point spread function of width λ/d and a residual seeing function of width λ/r_o where r_o is the Fried's parameter. Fried's parameter is a measure of the largest aperture which is diffraction limited for given seeing conditions. The fraction of power in the diffracted power is called the Strehl Ratio. Beckers estimates the range of Strehl Ratios in the K band to be 0.76 to 0.31. The J and H bands are significantly worse since r_o depends on wavelength as $\lambda^{5/6}$.

It is the large amount of power (on the order of 40%) in the broad seeing profile that makes ground based telescopes unacceptable for most of the investigations proposed in section 2 . The seeing profile is similar to the broad power distribution found in the aberrated HST mirror except that it varies on fast time scales as the seeing varies. Even in the stable case the current HST photometry is only good to 20% even with extensive computer enhancement. This sort of accuracy is unacceptable in the conduct of the distance scale programme. This extended wing also makes accurate resolution of sources such as shown in Figures 1 and 2 impossible.

Figures 10 through 12 show the encircled energy vs angular distance from the image centre for the line profiles discussed above. The Fried's parameters are taken

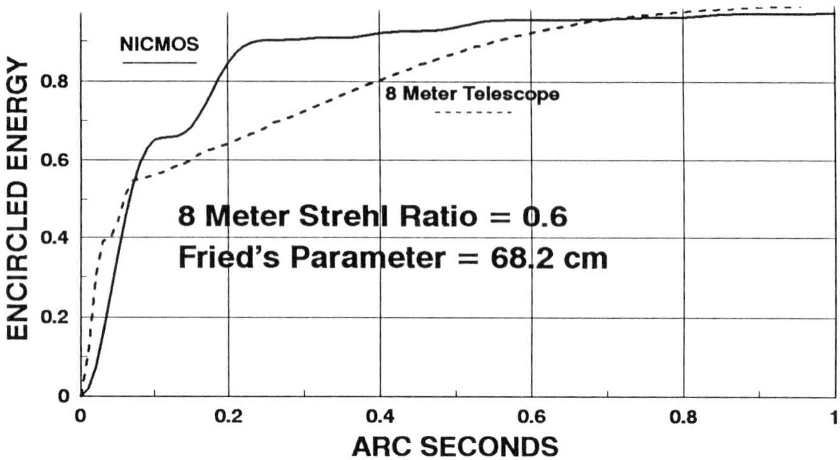

Fig. 10. J band encircled energy vs angular distance.

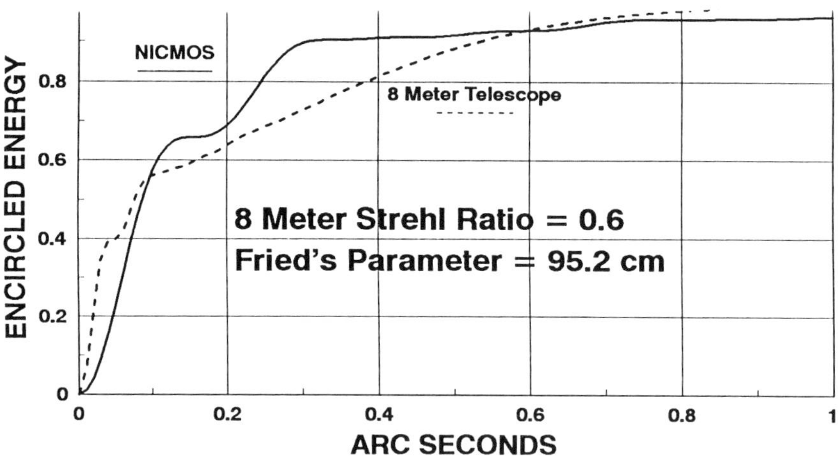

Fig. 11. H band encircled energy vs angular distance.

from the 'very good' line of Table IV and the Strehl Ratio is 0.6. These figures demonstrate the very large excess of energy in the 0.1 to 0.6 arc second region for the ground 8 metre over NICMOS profiles. It is this energy, which is variable with seeing, that makes interpretation of confused fields extremely difficult from the ground. It also demonstrates the limitations of high spatial resolution photometry from the ground.

Achievement of Strehl Ratios on the order of 60% requires corrections on the

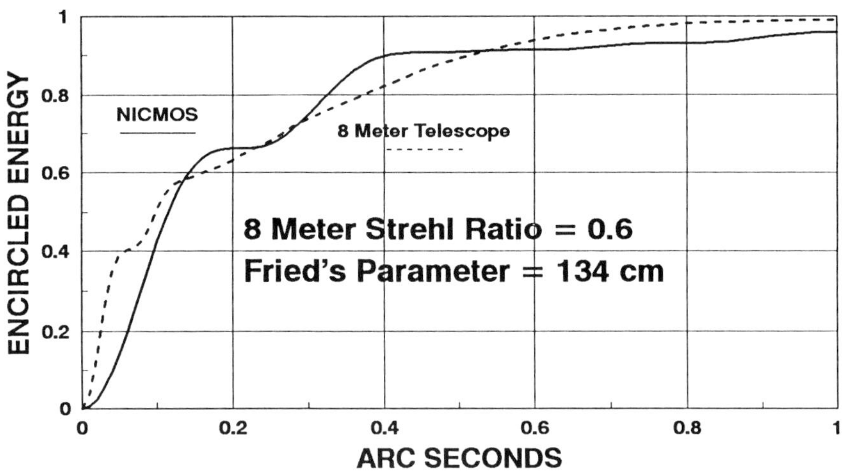

Fig. 12. K band encircled energy vs angular distance.

TABLE IV
Variations of D/r_o

		J		H		K	
Seeing quality	r_o at 0.5 μm	r_o	D/r_o	r_o	D/r_o	r_o	D/r_o
Mauna Kea average	14.4 cm	43.2	18.5	60.3	13.2	67.7	11.8
Good	15.5 cm	46.5	17.2	64.9	12.3	91.7	8.7
Very good	22.7 cm	68.2	11.7	95.2	8.4	134.	5.9
Excellent	25.0 cm	75.1	10.6	105.	7.6	148.	5.4

D = telescope diameter = 8 metres; $r_o \propto \lambda^{5/6}$; r_o = Fried's parameter.
For a given r_o the minimum image size is for $D/r_o = 8$.

order of 10 Zernike terms. Roddier (1991) indicates that this is only possible when the ratio of the telescope diameter to r_o is on the order of 8 or less and degrades rapidly for ratios larger than this. Table IV gives the Fried's parameters and ratios to telescope diameter for conditions measured on Mauna Kea using the data of Roddier (1991). It is easy to see that the appropriate conditions are usually not present even for the K band and only in the rarest of occasions present in the H or J band.

TABLE V
Imaging

Parameter	Camera 1	Camera 2	Camera 3
Pixel Size (arc seconds)	0.043	0.075	0.2
Field of View (arc sec x arc sec)	11 x 11	19 x 19	51 x 51
Diffraction Limited Wavelength	1 μm	1.75 μm	–

4. Scientific Instrument

NICMOS contains three cameras at different spatial resolutions which view different portions of the image plane. All of the cameras can therefore operate simultaneously. The cameras share the same fore or stage 1 optics which includes a three position camera selection mirror to move objects to different cameras without requiring HST repointing. This mirror is also at the first pupil of the instrument and is the place to introduce the optical figure to correct the HST mirror aberrations. Each of the cameras has a 256 $\times$ 256 HgCdTe detector array inside a cryogenic dewar at the final focal plane. The stage 1 optics focus the HST focal plane on field splitting mirrors which send the appropriate fields to individual optical systems for each camera. The individual optical systems determine the magnification for each of the cameras. Camera 2 has a coronagraphic mask and its filter wheel contains polarizers for polarization studies.

The NICMOS cameras have diffraction limited spatial resolutions at 1.0 μm and 1.75 μm to insure diffraction limited imaging at any wavelength in the 1.0 - 2.5 μm imaging spectral range. The third wide field camera provides high spatial resolution images over a wider field for survey work and extended images such as galaxies and planets. All of the optical elements of the cameras are outside of the dewar for easy alignment and access. Thermal control maintains the camera optics at a maximum stable temperature of 2°C to reduce thermal background emission and variation. The cryogenic dewar contains the second pupil apertures, filter wheels and detector arrays for all of the cameras. The cold filter wheels contain the polarization analyzers for camera 2. Table V shows the main characteristics of the cameras.

The detectors are 256 $\times$ 256 pixel hybrid arrays of HgCdTe photodiodes indium bump bonded to silicon direct readout multiplexers. All of the arrays have long wavelength cutoffs of 2.5 μm. The Rockwell International Corporation, under contract and in conjunction with the University of Arizona, developed these detectors specifically for the low signal, low background, and long integration time

TABLE VI
Detectors

Wavelength Range	$1 - 2.5\ \mu m$
Format (pixels x pixels)	256×256
Pixel Size (microns)	40
Read Noise (electrons)	20
Dark Current (electrons/sec.)	< 1
Linear Region (integrated electrons)	200 000

conditions required by NICMOS. Table VI gives the very excellent characteristics of the detector arrays.

4.1. Instrument Design

NICMOS consists of three major subsystems: the instrument/optical subsystem, the thermal and cryogenic subsystem, and the electrical subsystem. A mechanical structure supports and ties these systems together. The mechanical structure also provides the physical interface with the HST and properly locates the NICMOS optical axis relative to the HST optical axis. Each of the subsystems interacts with each other and the mechanical structure as well as the HST.

4.1.1. Optical System Design

Concepts were developed for each of the cameras and spectrographs by means of the ACCOS V lens-design programme of Scientific Calculations, Inc. The design goal, that the geometrical-optical aberrations should be smaller in diameter than 40 microns, is met in all cameras.

4.1.1.1. Cameras.

Stage 1. The cameras have common first stage optics (Stage 1) and separate magnification optics (Stage 2). Both stages are described below.

The three cameras share an initial imaging stage (henceforth called Stage 1), which reimages the telescope image plane in a more accessible area. It also provides camera selection and partial correction of telescope astigmatism. The reimaging mirror is a spherical toroid with unit magnification. The camera selection mirror is placed at an image of the telescope pupil and carries the correction for the HST spherical aberration.

Stage 2. In all three cameras the Stage-2 optics consists of two off-axis paraboloids. One has its focus at the centre of the field mirror, the other at the centre of the detector array. This arrangement allows a wide latitude in the location of the detectors. The detectors for each of the cameras are inside the dewar. There are no optical elements in the dewar.

The Stage-2 optics are coupled to the Stage-1 optics by field mirrors at the Stage-1 image plane. These have optical power to control pupil imaging. In Cameras 1 and 2 and 3 the pupil is imaged at cold stops in the dewar wall. These optics define the actual beam diameter of the cameras.

Modes of Operation and Image Quality. In the prime mode of observation the centre P of the Stage 1 field is imaged at the centre of the camera field in use. Both Stage-1 and Stage-2 are perfectly stigmatic at these points. Consequently, the aberrations in the field are minimal. Axial position and tilt of the image plane have been adjusted for best focus at the midpoints of the edges of the field. Due to field curvature, the centre is slightly defocussed. All aberrations are within the camera tolerances for diffraction limited imaging.

Camera Alignment. The optical system is focused and aligned without recycling the dewar. In each camera the field and the pupil are focused simultaneously by combined axial displacements of the Stage-2 relay mirrors. Similarly, the field and the pupil are aligned laterally by tilt and displacement of the folding mirror below the dewar.

4.1.2. Mechanisms

Virtual elimination of cryogenic mechanisms is one the significant aspects of the NICMOS design. The only moving cold parts are the filter wheels which do not require extremely precise motion. The mechanisms for moving them are, however, warm and mounted outside of the dewar. The following gives a brief description of the NICMOS mechanisms.

4.1.2.1. Camera Selection Mirror. The camera selection mirror is a multi position mechanism to move the NICMOS optical axis. The emphasis in this mechanism is on repeatability rather than absolute accuracy. The camera fields of view are oriented to make the mirror motion as simple as possible.

4.1.2.2. Filter Wheels. Each camera has a separate filter wheel, of similar design. The filter wheels mount on the optical access ports of the dewar. All cameras have a cold shutter portion for thermal blockage of the dewar opening during non-operational times. Warm motors and gears, mounted outside the dewar, drive the filter wheels so that no heat is dissipated in the dewar. The bearings and shaft are coupled with a bellows coupling to mitigate any warm to cold misalignment.

4.2. Cryogenics

The NICMOS cryogenic system consists of an inner tank of solid nitrogen surrounded by thermo-electrically cooled shields. The operating temperature of the inner cavity containing the detectors is 58 K. The nominal lifetime of the cryogenic system is 5 years operating time on-orbit with a 10% margin. The system is precooled to 40 K before launch which allows approximately 60 days of lockup before venting or servicing is required. This feature greatly eases launch handling.

The inner cavity is a vacuum chamber, sealed from the rest of the instrument and the telescope environment. Light enters through sapphire windows for each of the cameras. In this system no contaminants can reach the detector systems or any cryogenically cold surface. This prevents the difficulties encountered in the present instruments such as the WFPC. The only other components in the cryogenic cavity are the filter wheels as described in section 4.1.2. The cryogenic system is completely passive and is similar to other solid cryogen systems successfully flown by Ball Aerospace.

References

Beckers, J.M.: 1988, *ESO Conference on Very Large Telescopes and their Instrumentation*, Vol II. Ed. Ulrich, M.-H. Pub. European Southern Observatory.

Chokshi, A. and Wright, E.L.: 1988, *Astrophysical Journal*, **333**, 491.

Chokshi, A. and Wright, E.L.: 1989, *Astrophysical Journal*, **347**, 16.

Cowie, L.L., Gardner, J.P., Lilly, S.J., and McLean, I.: 1990, *Astrophysical Journal Letters*, **360**, L1.

Roddier, F.: 1991, *Publications of the Astronomical Society of the Pacific*, preprint.

Yoshi, Y. and Takahara, F.: 1988, *Astrophysical Journal*, **326**, 1.

SIRTF: SPACE INFRARED TELESCOPE FACILITY

E. F. ERICKSON and M. W. WERNER
NASA Ames Research Center, Moffett Field, CA 94035–1000, U.S.A.

(Received 24 June, 1991)

1. Programme

SIRTF will be a one metre class, cryogenically cooled observatory for infrared astronomy. It will achieve a 100- to 10,000-fold increase in sensitivity relative to previous facilities for observations between 3 and 700 microns. Three focal-plane instruments will provide photometric, imaging, and spectroscopic capabilities using large-format detector arrays. A five year lifetime in a 100,000 km orbit is planned. On-target efficiency should approach 85%, and roughly 85% of the time will be devoted to General Observer use. SIRTF will build on the successes of IRAS and COBE and on those to be obtained with ISO. SIRTF will complete NASA's planned family of Great Observatories. Table I compares anticipated capabilities of SIRTF with those of IRAS.

2. Science Instruments

Three science instruments will share the telescope focal plane. Light will be diverted to each of the instruments simultaneously by a pyramidal tertiary mirror. The annular field around the tertiary mirror is viewed by guidance sensors which control the secondary mirror tilt to achieve fine pointing. The three instruments are:

- IRAC: Infrared Array Camera
  Principal Investigator: G. Fazio, Smithsonian Astrophysical Observatory.
  Characteristics: Wide-field and diffraction limited imaging. 1.8–30 microns, using arrays with up to 256 × 256 pixels. Polarimetric capability. Broad and narrow filters.
- IRS: Infrared Spectrograph
  Principal Investigator: J. Houck, Cornell University.
  Characteristics: Long slit and echelle spectrographs. 2.5–200 microns using detector arrays with up 256 × 256 pixels. Resolving power from 100 to over 2000.
- MIPS: Multiband Imaging Photometer
  Principal Investigator: G. Rieke, University of Arizona.
  Characteristics: Background-limited imaging and photometry. 30–200 microns, with pixels sized for complete sampling of Airy disk. Array sizes up to

Space Science Reviews **61:** 95–98, 1992.

TABLE I
SIRTF/IRAS comparison

	SIRTF	IRAS
Primary mirror diameter	> 90 cm	60 cm
Telescope configuration	Cassegrain	Cassegrain
Cooling	liquid helium	liquid helium
Wavelength coverage	1.8–1200 μm	8–120 μm
Diffraction-limited wavelength	3 μm	$\sim 15\mu$m
Angular resolution	Diffraction limited at $\lambda > 4$ μm	15 arcsec at 12 μm
Pointing stability/accuracy	articulated secondary	scan
Field of view	7 arcmin	1 degree
Sensitivity*		
10 μm	6 μJy	70 mJy
60 μm	100 μJy	70 mJy
Number of detectors	$> 200{,}000$	~ 60
Spectral resolving power	> 2000	20
Mode	observatory	survey
Lifetime	5 years	10 months
Orbit	100,000 km circular	900 km, polar sun-synchronous

* For SIRTF, sensitivity is one sigma in 500 sec; for IRAS, it is one sigma per survey scan.

32 $\times$ 32 pixels. Broadband photometry, 200 to $>$ 700 microns. Polarimetric capability.

3. Science

The key scientific areas and capabilities are:

- Formation and evolution of galaxies – detection of protogalaxies to redshift > 5
- Ultraluminous galaxies and QSO's – spectroscopy of QSO's to redshift > 5
- Star formation – imaging and spectroscopy of low mass protostars throughout the galaxy
- Synthesis of the heavy elements – nickel, cobalt abundances in Virgo cluster supernovae
- Brown dwarfs – explore all regions of mass, age plane
- Search for other solar systems – detection of Vega-type debris disks around tens of thousands of stars; solar-type zodiacal cloud around closest stars.
- Solar system science – search for Kuiper belt comets
- Infrared and submillimeter background – extend COBE search for extragalactic infrared background

Fig. 1. Artist's impression of SIRTF in orbit.

– What lies beyond? – imaging surveys to < 0.001 of IRAS levels; first complete spectral surveys in infrared.

4. Recent Highlights and Status

SIRTF was selected as the highest priority new major mission for the 1990's by the Astronomy and Astrophysics Survey Committee (Bahcall Committee). This report describes SIRTF as follows: "SIRTF has superb science potential, is technically well developed, and has been preceded by two smaller and successful precursor missions."

Current plans call for starting development in 1994, with launch in 2000 or 2001.

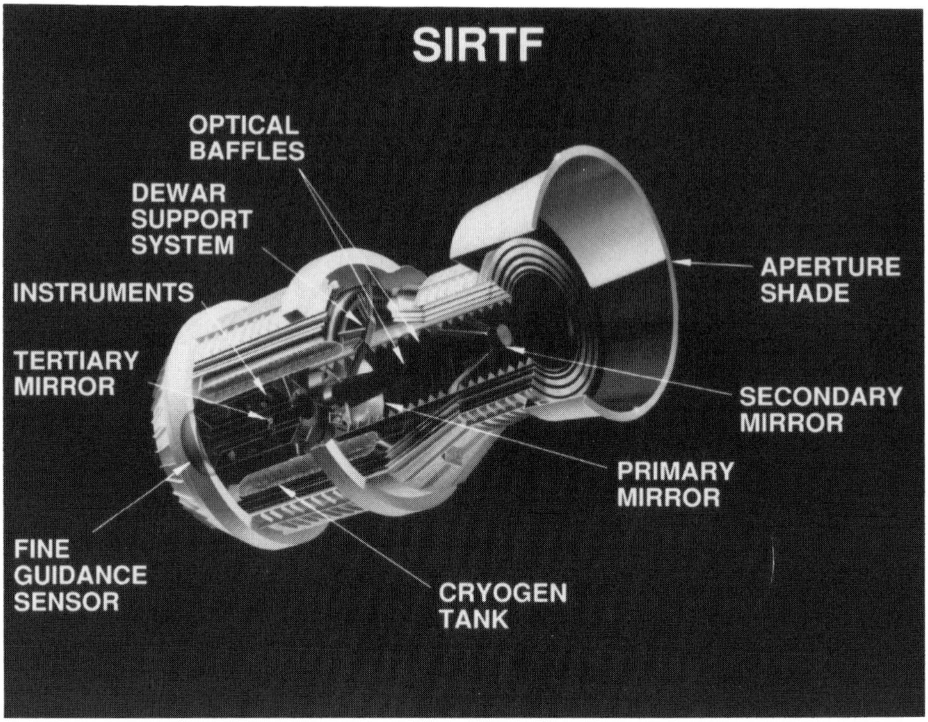

Fig. 2. Cut-away diagram of SIRTF.

Reference

Werner, M. W. and Eisenhardt, P. G.: 1988, *Astophys. Lett. and Comm.* **27**, 89.

PLANS FOR SUBMILLIMETRE AND INFRARED SATELLITES IN JAPAN

M. HAYASHI
Department of Astronomy, University of Tokyo, Bunkyo-ku, Tokyo 113

(Received 13 June. 1991)

Abstract. I describe plans for submillimetre and infrared satellites which are now under discussion among Japanese infrared and millimetre astronomers. We propose a submillimetre and/or infrared mission to be launched around 2000 with a main reflector $\sim$1–2 m in diameter and cooled down to 50–150 K. This mission, although a small one, will be able to contribute significantly to the advance in submillimetre and infrared astronomy if its role is specified for suitable purposes like surveys.

Key words: space mission, infrared astronomy, submillimetre astronomy.

1. Organization

The number of scientists in the space submillimetre and infrared programme in Japan is much smaller than that in the US or in Europe. Planning for the future submillimetre and infrared mission is still in the early phases, but there are at least the following groups of infrared and millimetre astronomers who try to organize a bigger group for the space mission.

Institute for Space and Astronautical Sciences (ISAS) is the centre of space astronomy, and Prof. Okuda leads a group of several infrared astronomers there. There are also a number of X-ray astronomers and rocket and/or satellite engineering scientists in ISAS. Prof. Matsumoto has a group of several infrared astronomers at *Nagoya University*. Those two groups of infrared astronomers together with collaborators in domestic and foreign institutes are launching an infrared space telescope (IRTS) as their first space infrared mission. The IRTS project is now scheduled for launch in 1993 and is in the final stage. Some of the radio astronomers in Japan especially those belonging to the *Nobeyama Radio Observatory* (NRO) and to the *University of Tokyo* wish to expand their frequency coverage up to the submillimetre region through experiments from space, with balloons, or from Antarctica. Future space plans for submillimetre and infrared astronomy in Japan are now being discussed among these groups.

2. Space Development

There are two space developing institutes in Japan, ISAS and NASDA (*NAtional Space Developing Agency of Japan*). I will briefly mention here the basic difference between these two space institutes.

Space Science Reviews **61:** 99–102, 1992.

ISAS is a relatively small institute which belongs to the Ministry of Education, Science and Culture. Its annual budget is limited to ~\$M 130, and ISAS has only small launchers. However it has an important role in launching satellites, mainly for scientific purposes, at a rate of almost one mission every year. It now uses the rocket M3S-II, which can launch a 500 kg satellite up to the Low Earth Orbit (LEO). ISAS is now developing a slightly larger rocket M-V which will be available from 1994. M-V has an ability to launch a 2 ton satellite up to LEO. However the efficiency factor for scientific instruments is not as high as that for US and European satellites, and we can only utilize 600 kg for the weight of scientific instruments out of 2 tons of the satellite mass at LEO.

The other institute for space development in Japan is NASDA which is related to the Ministry of Industrial Sciences and Engineerings. The annual budget for NASDA is ~\$M 1,000. Commercial and engineering test satellites are usually launched by NASDA with much larger launchers compared to ISAS. NASDA presently uses the H-I rocket which has an ability of launching ~500 kg of total mass up to the High Earth Orbit (HEO). NASDA is developing a larger rocket called H-II, which is capable of launching a 2 ton satellite up to HEO. At present it is difficult for Japanese astronomers to use NASDA's rockets for launching satellites exclusively for astronomical purposes partly because of the budgetary limitations.

3. Scheduled Missions

Scheduled scientific missions for infrared and submillimetre astronomy around 2000 are shown in Figure 1 together with the launching schedule of ISAS scientific missions. In the same year 1993, when ISO is to be launched, we have an infrared mission IRTS (*InfraRed Telescope in Space*), which is a liquid He cooled infrared telescope of 15 cm in diameter mounted on the *Space Flyer Unit* (SFU) to be launched by the NASDA's H-II rocket. SFU is a kind of space experiment module and IRTS must share the space with other instruments, which limits the total size of the telescope. During its lifetime of 3 weeks IRTS will measure diffuse emission with 4 detectors, namely with NIRS (1.3–4.1 μm), MIRS (4.3–1.6 μm), FILM ([CII], [OI], etc.), and FIRP (15–80 cm^{-1}).

Many ISAS missions are scheduled in 1990s for the astronomical, planetary, and geophysical purposes as shown in Figure 1. SOLAR-A (which was launched successfully in August 1991 and has been renamed as the YOHKOH mission) and ASTRO-D are X-ray satellites and VSOP is a space VLBI observatory to be launched in 1994 with the M-V rocket. X-ray, lunar, and planetary missions are scheduled or will be scheduled in late 1990's. Mission schedules after 1998 are not rigidly determined and we will have a chance to launch a submillimetre and/or infrared telescope around 2000.

The next generation infrared and submillimetre space projects in the US and in Europe are to launch large, observatory-type missions such as EDISON, FIRST, SIRTF, and SMMM at the beginning of the 21st century. Figure 1 shows that

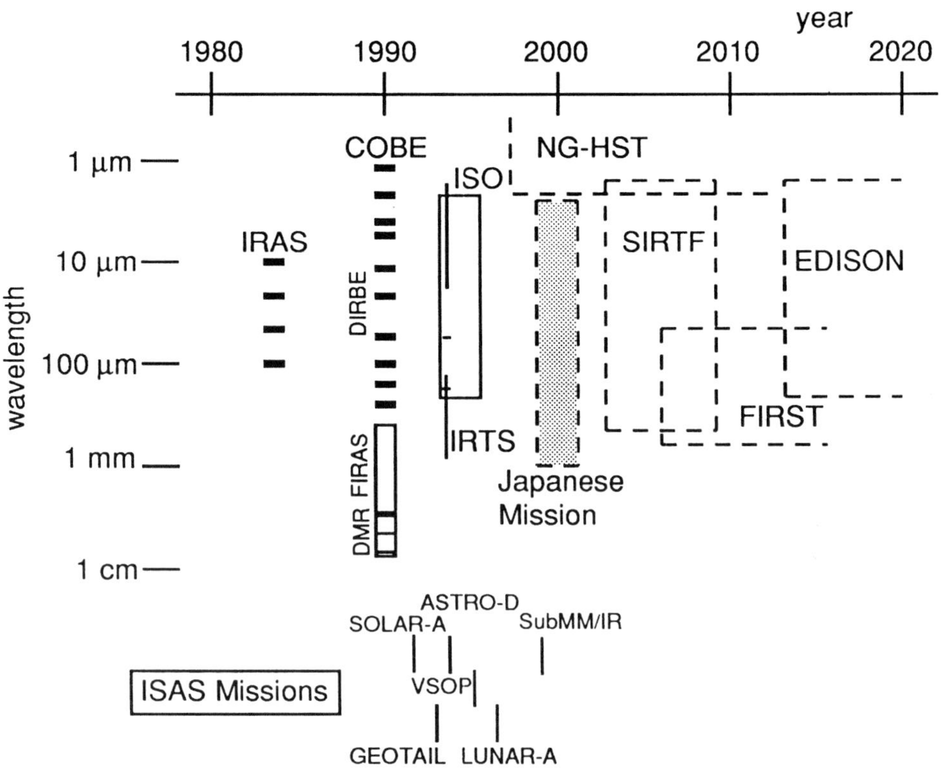

Fig. 1. Space infrared and submillimetre missions.

we expect no mission covering mid-infrared to submillimetre wavelengths around 2000. A small ISAS mission before these larger ones will have great importance if it is dedicated to suitable scientific purposes like surveys.

4. Plans for a Submillimetre and Infrared Satellite

As I have been mentioned above, Japanese infrared and submillimetre astronomers have severe constraints on space programmes. Although it seems difficult for us to launch a large mission with NASDA's H-II rocket around 2000, ISAS M-V rockets will be available for submillimetre and infrared missions around 2000. With the ISAS M-V rocket the diameter of the telescope is limited to <2 m and the total weight for scientific instruments is <600 kg for LEO.

The feasibility and scientific importance of Japanese infrared and submillimetre satellites are being discussed among the satellite working group for infrared astronomy. One such working idea is SMIRT (*SubMillimetre and InfraRed Telescope*), which is an infrared telescope of 1.5 m in diameter cooled down to 50 K by radiative and mechanical cooling. SMIRT, proposed by Prof. Okuda, covers the whole infrared and submillimetre wavelength range with NIR/MIR/FIR observations and bolometer arrays for submillimetre observations. Although the natural background limit is achievable only in the NIR/MIR wavelength, FIR and submillimetre observations are also valuable because of the reduced confusion limit due to the large telescope diameter. Prof. Matsumoto planned an infrared survey satellite mainly dedicated to the 10 and 20 μm wavelength observations with array detectors on the similarly cooled telescope with smaller aperture. Radio astronomers proposed another mission whose main target is in the submillimetre region, to be observed with a radiatively cooled 2 m telescope with bolometer arrays and heterodyne receivers. In any case an observatory-type mission seems rather difficult mainly due to the difficulty of its attitude control at LEO. I think a survey mission like IRAS or ROSAT will be very powerful for submillimetre and infrared astronomy if its spatial resolution and sensitivity are higher than the previous missions, even if our satellite is small and short-lived.

NASA'S FUTURE PLANS FOR SPACE ASTRONOMY AND ASTROPHYSICS

MIKE KAPLAN
Chief, Advanced Programs Branch, Astrophysics Division, Office of Space Science and Applications, NASA Headquarters, Washington, DC. 20546

1. Future Mission Plans

Based on the recommendations of *The Decade of Discovery in Astronomy and Astrophysics* – a report from the National Research Council's Astronomy and Astrophysics Survey Committee (chaired by John Bahcall) – NASA's Astrophysics Division crafted a strategic plan consisting of missions and other programme elements to begin development within the next ten years. NASA's challenge in developing this ten-year plan was to implement the Bahcall report recommendations within the very constrained fiscal climate. The resulting strategic plan was approved by the science community and presented to the Space Science and Applications Advisory Committee (SSAAC), NASA's highest level science advisory group, at a week long strategic planning retreat. SSAAC combined the Astrophysics plan with the needs of the rest of the space science community and recommended a single, fiscally responsible plan for NASA space science missions from 1994 through 1998.

Along with future missions, the space science strategic plan identified enhancements to NASA's research and technology base to support the missions. Astrophysics missions in the future mission queue are given below, along with brief descriptions. The accompanying dates indicate the planned fiscal year for commencing development activities. For missions initiated after 1998, the commencement years are used for technology planning purposes only.

1.1. The Space Infrared Telescope Facility (SIRTF) - 1995

SIRTF is a one metre-class, cryogenically-cooled infrared observatory in high earth orbit (100,000 km). Its state-of-the-art detectors provide a 100 to 10,000 gain in sensitivity over its predecessors and also provide up to a million-fold improvement in imaging capability due to the increase in sizes of arrays in the near and mid-infrared, which typically consist of tens of thousands of individual detector elements. The continuum and spectral line imaging capability between 2 and 200 microns and the photometric capability at longer wavelengths, allow a variety of significant astrophysical problems to be investigated in unprecedented detail, including the formation of planets, solar systems, and the earliest evolutionary

Space Science Reviews **61:** 103–112, 1992.

stages of galaxies. Over 1,000 research groups will have access to SIRTF during its 5 year lifetime. SIRTF, which will complete the Great Observatories, was identified in the Bahcall report as the highest priority new initiative in all of astronomy in the 1990s.

1.2. The Stratospheric Observatory for Infrared Astronomy (SOFIA) - 1995

SOFIA is a 2.5 m class airborne observatory based on a modified Boeing 747SP aircraft. It will succeed the highly successful Kuiper Airborne Observatory (KAO). SOFIA will provide three times the angular resolution and ten times the sensitivity of the KAO over the entire wavelength range from 0.3–1,600 microns. During its anticipated 20+ year lifetime, up to 4,000 research flights will be flown in support of forefront research and to train the next generations of space infrared/submillimetre experimentalists and observers. SOFIA will offer high spatial and spectral resolution as a complement to SIRTF's exquisite sensitivity and imaging capability. SOFIA will concentrate on studies of the brighter objects and events in the nearby universe, where the major astrophysical issues require its higher spatial and spectral resolving power. With SOFIA's frequent flights and 'hands-on' environment, instruments do not have to be highly automated and can be developed and refined relatively quickly. This allows students the opportunity to participate in research projects from start to finish within the period of their graduate studies and also allows for the incorporation of state-of-the-art detector systems and spectrometers.

1.3. The Submillimetre Intermediate Mission (SMIM) - 1998

SMIM is a passively-cooled, 2.5 metre space observatory in a 70,000 km elliptical orbit for studying the 100 to 750 micron spectral region. The principal objective of this mission is to obtain a complete, high-resolution (up to 0.1 km/sec) spectral line survey of an estimated 100 astronomical sources or positions during its one year lifetime. The submillimetre spectral region contains clues to the processes by which diffuse gas clouds are assembled into stars and planetary systems. Besides measuring the thermodynamic properties, SMIM will provide a detailed chemical analysis of star and planet forming clouds. SMIM will greatly extend the SIRTF and SOFIA submillimetre capabilities by studying many lines of each of hundreds of molecular and atomic species at extremely high spatial and spectral resolution.

1.4. The Astrometric Interferometry Mission (AIM) - 1999

Planned for launch in the early 21st century, AIM will be the first optical interferometer mission in space. It will be capable of spatial resolution of 3–30 microarcsecond of 20th magnitude objects – outperforming any current or planned optical telescope, and exceeding that achieved by very-long-baseline arrays of radio telescopes using analogous interferometric techniques. AIM will make highly accurate measurements of the positions and motions of distant stars, star clusters, quasars, and other faint astronomical objects. These fundamental new data will furnish precise distances to objects throughout our galaxy, vastly improving the calibration of the cosmic distance scale and permitting definitive estimates of the size and age of the Universe. Observations of the dynamics of globular star clusters in our galaxy will shed new light on galactic formation and evolution. AIM also will be capable of detecting planets around remote stars by measuring perturbations in stellar motions caused by orbiting companions. Two promising candidate designs for AIM are already under study. Both make use of Michelson interferometers, and both can achieve the mission objectives; a choice between the two designs will be made in 1993. A Space Interferometry Science Working Group (SISWG) has been established to define and guide the AIM programme.

1.5. THE GREATER OBSERVATORIES - 2001

These will succeed the current Great Observatories, i.e. Hubble Space Telescope, Gamma Ray Observatory, Advanced X-ray Astrophysics Facility and SIRTF. Based on the recommendations of the Bahcall report, there are many potential candidates for the Greater Observatories programme, including

- Next generation space telescope – a 6 to 8m class instrument covering the near IR, visible and UV wavelength regimes,
- Laser gravitational wave observatory,
- Infrared and optical imaging interferometer,
- Passively-cooled infrared observatory with a 2.5m class aperture,
- Large X-ray imaging and spectroscopic observatory,
- Submillimetre observatory with either a 10m class aperture or an interferometer, and
- Radio telescopes in space consisting either of a large single aperture or a space-based array.

The priorities for the Greater Observatories programme will be set by careful peer review conducted by the entire space astrophysics community and will be based on the importance of the science and the readiness of the technology.

1.6. MISSION FROM PLANET EARTH (MFPE)- START DATE UNDETERMINED

The Astrophysics MFPE programme will pursue mission concepts that have the potential to perform unique science from the Moon: science that would be infeasible or inefficient from any other location, including Earth orbit. The Astrophysics MFPE programme may also take advantage of man's return to the Moon for non-Astrophysics purposes such as to study the moon itself or to act as a way station for Mars exploration. A permanent lunar base may be an enormous windfall to Astrophysics since manned assembly, and possibly even manufacturing, of lunar observatories could enable missions that would not otherwise be possible. Potential missions include optical and submillimetre interferometers with hundreds of metre baselines, and imaging X-ray telescopes with kilometre-long separations between mask and detector.

Since the MFPE began in July 1989, there have been several workshops and reports including the '90-Day Study' and the NASA sponsored conference on Astrophysics from the Moon (1990). These reports document studies of potential lunar observatories, make recommendations on missions, and help identify the technology that enables these missions. There are five mission concepts that have been recommended for lunar siting and are attractive because of unique advantages of the Moon:

- Lunar Transit Telescope (LTT),
- Lunar very low frequency radio array,
- Lunar optical interferometer,
- Lunar submillimetre interferometer, and
- Large lunar telescope.

To determine if and when the missions will be developed, and whether they are best sited on the moon, each mission will be evaluated with respect to science capability, reliability, expense and availability (technological readiness).

One mission is worthy of particular attention: the Lunar Transit Telescope (LTT). Several studies have identified this mission as the best first observatory on the Moon. It requires little technology development, is low risk in that it lacks moving parts, has an earth prototype already in operation, is among the lowest cost proposed Astrophysics lunar mission, is light enough to be launched from Earth as a single unit with lander, and can be delivered and operated autonomously. It also would be an excellent environmental test-bed for later, more expensive, lunar missions. The detector technology is already being developed to address the disadvantages of lunar siting such as the radiation environment. Combined with an extraordinary science return, these characteristics appear to make the LTT a logical choice for the first lunar science mission.

2. Future Technology Needs

NASA recognizes that implementing the missions described above will involve many technical challenges. The Astrotech 21 programme was created to develop a strategy and plan for the development of the required technology. Under its auspices, several workshops have been conducted to identify specific technology needs, e.g. information technology, sensors, and optics technology. These results will then be synthesized into an overall Astrotech 21 Technology Development Plan. This plan will then be implemented jointly by the Astrophysics Division and NASA's Office of Aeronautics and Exploration Technology (OAET).

Based on preliminary results, NASA will need an aggressive technology development programme to meet the technological challenges posed by each future space astrophysics mission. NASA has a sensors technology programme in place, but it is meeting only a portion of the technology development needs for future space astrophysics missions and needs to be augmented. NASA must also build a strong commitment to develop optics technology for future space astrophysics missions.

Below is a brief description of technology development needs for future space astrophysics missions:

SIRTF: The greatest technology challenges facing SIRTF are in cryogenic optics and mechanisms, integration and testing, and infrared detectors. In each area, a technology development programme begun during the 1980s has established the feasibility of meeting the SIRTF requirements and is continuing with the demonstration of hardware prototypes.

- *Cryogenic Optics* – SIRTF requires a primary mirror one-metre in diameter that delivers an image diffraction-limited at 3 μm when cooled to below 4K. This is a step beyond what had been demonstrated by the Infrared Astronomy Satellite (IRAS). Since SIRTF's image quality requirement is routinely exceeded in modern optical practice, the technology challenges are in the cryogenic application. Tests at the NASA/Ames Research Center showed that a half-scale fused silica mirror can be thermally cycled from room temperature to 4K and predictably retain figure. The Infrared Space Observatory (ISO) team has reported successful cool-down and optical test at 4K of a 64-cm diameter fused silica mirror in a flight-like mount. This mirror, when tested cold, showed diffraction-limited performance at 3.5 μm, exceeding the ISO requirement of 5 μm, and very close to what is required by SIRTF on a slightly larger scale. JPL is designing a full-scale mirror – with mounts – that will be fabricated at half-scale and tested to demonstrate compatibility with the launch environment and that the mirror mounting does not adversely affect figure when the mirror is thermally cycled.
- *Cryogenic Mechanisms* – The SIRTF telescope and instruments incorporate many mechanisms that must work reliably at cryogenic temperatures. Hardware brassboards of the principal instrument mechanisms – filter wheels and grating drives – have been fabricated and tested extensively at cryogenic temperatures, and demonstrated performance consistent with a five-year lifetime. The principal mechanism in the telescope is the articulated secondary mirror, which must operate precisely at temperatures below 4K with high stability and low power dissipation. This is unique in that neither IRAS nor ISO incorporates an articulated secondary mirror.
- *Integration and Test* – A plan for test and verification of the performance of the instruments and the telescope at cryogenic temperatures is being developed and fed into the schedule for assembly and integration of the flight hardware. This programme will go from cryogenic shake testing of individual detector assemblies and mechanisms to verification of the optical performance of the full observatory. One promising approach under consideration for system test is to incorporate a plane mirror into

the aperture cover that seals the dewar. An alignment light inside one of the instruments could be projected through the entire optical system and off the plane mirror, and brought to a focus by the telescope on one of SIRTF's arrays. This would permit verification of instrument alignment and end-to-end system optical performance at the SIRTF operating temperature.

– *Infrared Detectors* – The most challenging technology requirement for SIRTF, and the one with the biggest payoff, lies in the detector arrays. Devices with much lower noise than is acceptable for ground-based application are required to take full advantage of the low-infrared background provided by SIRTF. Because of the very broad spectral coverage of SIRTF, the instruments incorporate six different types of detectors. At the time of selection of the SIRTF instruments, infrared arrays were not yet in common use for astronomy, and single detectors meeting SIRTF's performance requirements had not been demonstrated. Since then, a technology development programme aimed at the SIRTF requirements has been carried out jointly by the three instrument teams and other elements of the project, with contributions from other programmes within NASA and invaluable collaboration by commercial detector manufacturers.

Infrared arrays like those to be used on SIRTF consist of a detector plane mated to a multiplexer that provides an electronic interface to the outside world. The arrays will be operated in a staring mode in which electrical charge generated by absorbed infrared radiation is integrated and read out at intervals ranging from 1 to 1,000 seconds. The technology programme has focussed on both the detector materials and the multiplexers, and brought us to the point where devices meeting or exceeding the SIRTF requirements in the most fundamental parameters – noise, quantum efficiency, and dark current – have been met in all but one of SIRTF's basic detector types. The present status of each detector technology effort within the programme is summarized below.

$\lambda < 30\ \mu$m – The imaging and spectroscopic instruments will use InSb (2–5 μm) and Si:As (5–28 μm) detectors mated to silicon multiplexers. For the InSb arrays, the quantum efficiency and dark current meet the SIRTF requirements, and the noise is within a factor of three of the desired level. Programmes to achieve further noise reduction by optimization of the new low-temperature multiplexers described below and by implementation of sampling techniques demonstrated at ground-based observatories are under way. The Si:As arrays meet all SIRTF requirements. The principal technical challenge remaining in this wavelength range is to extend the demonstrated performance to arrays with the format required by SIRTF, about an order of magnitude more pixels than the arrays demonstrated to date. Two parallel programmes are underway to extend low-noise, low-temperature multiplexer technology to the format required by SIRTF, and the first tests of InSb and Si:As arrays at this scale (256×256 and 128×128 pixels, respectively) are now beginning.

$30\ \mu m < \lambda < 200\ \mu m$ – Several types of doped germanium photoconductors will be used for imaging and spectroscopic applications. Much smaller arrays (fewer pixels) are required at these longer wavelengths because of the increased angular size of the diffraction image. The detector plus readout performance required has been demonstrated in single detectors and linear arrays. The principal challenge is the extension of this development to the manufacture of modest-sized area arrays that meet flight-quality standards; proof-of-concept articles based on several approaches will be manufactured and evaluated within the next year.

$\lambda > 200\ \mu m$ – SIRTF will make use of several bolometers that will be cooled to 0.1K by an adiabatic demagnetization refrigerator similar to one being developed for flight on AXAF. A bolometer that is within a factor of two of meeting the SIRTF requirements has already been demonstrated in the laboratory, and a prototype refrigerator optimized for SIRTF use will be tested on a balloon flight within a year.

SMIM: There are four areas where SMIM has a baseline technology that is new or has not been space qualified: the SIS heterodyne radiometer, the composite aperture, an adiabatic demagnetization refrigerator, and an inflatable, space-rigidized sunshade.

- *SIS Heterodyne Radiometer* – SIS receivers are currently in use on ground-based telescopes for wavelengths longer than 600 μm. An OAET submillimetre sensors programme is in place that will demonstrate a 450 μm SIS receiver, subsequently to be incorporated in the Caltech Submillimeter Observatory (CSO). An augmentation to this programme has been proposed for FY 92 that will lead to the development of the components for a receiver at 350 μm. Further development is needed for a 250 μm receiver. These receiver systems will be used in astronomical applications and checked from an engineering viewpoint on the NASA airborne facilities KAO and SOFIA. As a baseline, SMIM has SIS heterodyne receivers to 250 μm, but alternatives exist, either to use Schottky diode receivers for the shortest vavelengths or to extend the range of the Fabry-Perot spectrometer.
- *Composite Aperture* – The Precision Segmented Reflector (PSR) programme currently plans to produce composite panels with the proper characteristics in the 1995 timeframe. Panels with the required radius of curvature, capable of operation down to 200 K, should be tested in FY 92 and panels for 100 K are planned for FY 94. Reflectors of ULE glass or silicon carbide could be substituted with cost and weight penalties.

- *Adiabatic Demagnetization Refrigerator* – Adiabatic demagnetization refrigerators are currently under development for both the Advanced X-Ray Astronomical Facility (AXAF) and the Space Infrared Telescope Facility (SIRTF).
- *Inflatable Sunshade* – The inflatable sunshade is being developed for the European Far Infrared and Submillimeter Space Telescope (FIRST) mission and should be demonstrated by 1992. Alternate designs exist at the cost of a slight reduction in the diameter of the primary reflector.

AIM: Studies of AIM options have revealed many critical technology areas for the spacecraft, including picometer metrology, active delay lines, and control of vibrations of large structures in space. These area have been provided by OSSA as inputs to the OAET technology programme, and the last area is already the subject of the OAET Control-Structures Interaction (CSI) programme.

- *Picometer Metrology* – Since astrometric accuracy is approximately equal to the ratio of the uncertainty in the baseline length to the baseline length, high accuracy metrology systems are required. For example, to achieve 10 mas (50 picorad) accuracy over a baseline length of 10 m requires an uncertainty in the baseline length of 500 picometers. Of course, for shorter baselines, the required metrology accuracy is tighter, and, depending on the magnitude of other sources contributing to the baseline-length error budget, the required accuracy in the metrology systems may be tighter still. If superior metrology accuracy can be achieved, it may make it possible to build an astrometric instrument of shorter baseline.
- *Active Delay Lines* – The active delay lines equalize and maintain the optical path lengths in the interferometer. The measurement of the position of the delay line mirror at which the path lengths are equal provides a key value in determining the angular position of the astrophysical target. Besides the important path measurement, disturbance to the optical path differences at frequencies up to a few hundred hertz are effectively taken out by servos controlling the lengths of the delay lines.
- *Control-Structures Interaction* – AIM requires extremely stable optical paths, and a multifaceted approach is used to reduce the vibrations to sufficiently low levels. First, the basic spacecraft design philosophy is to eliminate common sources of vibration whenever possible, such as a propulsion system (natural orbit decay during the lifetime of the mission is acceptable) and tape recorders (an all-solid-state system is used). Second, passive damping and isolation techniques are being incorporated into the design. It is expected that with this technology, available from the CSI programme, the spacecraft will be mechanically stable enough that the instrument can operate in the required quiet environment.

Greater Observatories: As was discussed earlier, future astrophysics missions require technology advancements beyond the state-of-the-art. Under Astrotech 21, a phased technology development programme plan would be implemented to meet future space astrophysics needs. Estimates of funding requirements to develop all the technology needed for future space astrophysics missions is roughly \$50 to \$75 million dollars per year over the next decade. Because of the enormity of this requirement, technology development will be undertaken jointly with OAET. Using outputs of the Astrotech 21 studies as a guide, the Astrophysics Division will work closely with OAET as they formulate their technology development plans. This coordination will ensure both the timely development of critical technologies and the smooth transition from basic technology efforts to flight programmes. Additional Astrophysics advanced technology development funding is required to be able to contribute to this critical technology development and to achieve the leverage made possible by joint OAET and Astrophysics Division technology development programmes.

MFPE: At present, MFPE is awaiting congressional approval. Meanwhile, the Astrophysics Division has begun supporting initial mission and technology studies at a low level through the advanced technology development programme and within the Astrotech 21 programme. The Bahcall report makes several recommendations regarding the strategy for a lunar-based astronomy programme. These recommendations, listed below, reflect the Division's strategy toward the MFPE programme.

- "The committee recommends that NASA should formulate its plans for a lunar initiative in a way that protects the base programme in astrophysics from disruptions caused by problems in the Space Exploration Initiative."
- "Site survey observations from the Lunar Observer(s), possibly with soft landed experiments such as a small transit telescope, should be a high priority for a lunar programme. The requirements for astronomical observations should be carefully considered in the selection of the site for a lunar base."
- "NASA should develop the technology necessary for constructing large telescopes and should investigate which of these facilities are best placed in earth orbit and which are best placed on the moon."

Acknowledgements

I wish to acknowledge the contributions of the SIRTF, SOFIA, SMIM and AIM project teams throughout the country. I also wish to thank the many scientists and engineers participating in the Astrotech 21 programme as well as the many NASA programme and project managers at NASA Headquarters as well as at the Goddard Space Flight Center, Ames Research Center, the Jet Propulsion Laboratory, and the Marshall Space Flight Center.

OPTIMISED RADIATIVE COOLING OF INFRARED SPACE TELESCOPES AND APPLICATIONS TO POSSIBLE MISSIONS

T. G. HAWARDEN

Joint Astronomy Centre, 665 Komohana Street, Hilo, Hawaii 96720, U.S.A. and Royal Observatory, Blackford Hill, Edinburgh EH9 3HJ, U.K.

R. O. CUMMINGS and C. M. TELESCO

NASA Marshall Space Flight Center, ED-63, Huntsville, Alabama 35812, U.S.A.

and

H. A. THRONSON, JR

Wyoming Infrared Observatory, Campus Box 3905, University of Wyoming, Laramie, WY 82071, U.S.A.

Abstract. General principles are outlined for the design of space infrared telescopes intended to cool by radiation to the lowest temperatures attainable without the use of on-board cryogens, and assuming on-orbit cooling after a warm launch. Maximum protection from solar and earth heating, maximum radiating area and efficiency and minimum absorbing area and absorptivity are the obvious basic criteria. The optimised design is a short, fat telescope surrounded by a series of radiation shields, each cooled by its own radiator. Maximising the longitudinal conductivity of the radiation shields and of the telescope tube itself is important both to the on-orbit cooling time and the final achieveable temperature. Realistic designs take between 80 and 200 days to cool to within a few degrees of equilibrium temperatures, depending on the materials used. Great advantages accrue from the use of an orbit distant from earth. Both simple models and detailed simulations suggest that temperatures of 30 to 40 K are attainable in high earth orbits. Placing a radiatively cooled telescope in a halo orbit around the Lagrangian point L2 is a particularly attractive option and significantly lower temperatures can be achieved there than in Earth orbit. Optimised radiative cooling is an important element of the small Japanese mission SMIRT. We suggest that a combination of an ESA Medium-sized Mission with a NASA Explorer to send a 2m+ telescope to an L2 halo orbit would provide a cost-effective and powerful long-duration facility for the early 21st century.

1. Introduction

The astronomical infrared spectral region conventionally covers wavelengths between about 1 micron and about 300 microns, where the submillimetre band commences. The advantages of space-based over ground-based infrared telescopes are well known and are summarised earlier in these Proceedings as well as in the literature (e.g. Davies *et al.*, 1991).

In addition to eliminating the deleterious effects of atmospheric absorption and emission in the infrared, a particularly important advantage of placing the telescope in space is the opportunity to reduce the temperature of the optical surfaces and associated structure without danger of their contamination by the condensation and freezing of water and other atmospheric gases. At typical earth-based temperatures, and even at those attainable in the Antarctic or in the stratosphere,

Space Science Reviews **61:** 113–144, 1992.

telescope structures and optical components emit significant radiation longwards of about 3 microns. Despite the most careful design this thermal emission makes the detection of faint celestial sources enormously more difficult than in the visible. Nevertheless, because most of the energy in the universe ends up in the infrared, the fullest possible exploitation of this band is essential to the development of modern astronomy, and the cooling of space infrared telescopes is consequently a technical issue of great importance for the science.

In a companion paper Thronson *et al.* (1993) have outlined the effects of telescope temperature on performance under a wide range of assumptions. In brief, photon noise from background radiation sets the limit to performance at any wavelength. The lowest level is eventually set by natural (astronomical) background emission at all wavelengths irrespective of telescope temperature. For all non-zero telescope temperatures, however, there is a transition wavelength, longward of which the natural background is dominated by the telescope thermal emission as the main factor limiting sensitivity. In the vicinity of the earth, when looking in directions in which the natural background surface brightness is lowest, emission from an optimally designed telescope at a temperature of 100 K begins to dominate that from the background at ~ 7 microns. At 40 K the transition occurs at ~ 20 microns and at 10 K around 100 microns. Longward of the critical wavelength other factors play a part in determining the relative performance of a particular telescope. Thus, a larger collecting area will keep the performance of a big telescope ahead of that of a smaller, colder one to somewhat longer wavelengths, while the use of small spectral bandpasses – e.g. in spectroscopy – can also favour a larger, warmer instrument in many circumstances when the limitations of real detectors are taken into account.

Confusion by background sources (galaxies and the IR "cirrus") may become important at wavelengths well below 100 microns (Helou and Beichman, 1990; Gautier *et al.* 1992). Because the angular resolution of the telescope is set by diffraction, the performance of a telescope for broadband deep imaging then becomes heavily dependent on its diameter, so that the raw sensitivity advantage of a very cold telescope may be nullified by the larger aperture of a warmer instrument. Finally, at wavelengths longer than about 150 microns (on the Rayleigh-Jeans side of the spectrum of the telescope background emission) the temperature of the telescope begins to play a much less important rôle in its sensitivity than does its collecting area. Telescopes designed for the submillimetre regime, therefore, again benefit much less from low temperatures than from large aperture.

Two main techniques are available for cooling a telescope in space: cooling by liquid cryogen (usually helium) and cooling by radiation. The former has been and is being used in the first generation of infrared space telescopes, to achieve temperatures of a few kelvin. The latter has been proposed for several large-diameter submillimetre telescopes and explored for large general purpose missions. Most of these studies predict equilibrium temperatures in the range from 150 to about 100 K. The two techniques have thus tended to be seen as appropriate for quite distinct regimes and applications, although several infrared experiments on the recent cryogenic Cosmic Background Explorer mission (COBE) are continuing to function well at temperatures around 70 K maintained by radiative cooling after the exhaustion of the liquid helium cryogen load.

The two main cooling techniques have well-defined advantages and disadvantages. Cryogenic cooling can readily provide temperatures so low that telescope emission becomes unimportant throughout the infrared, while the cryogens make it simple to maintain the instruments, and most importantly their detectors, at temperatures low enough both to permit them to function and to prevent them from being saturated by background radiation from their enclosures. However reliance upon cryogens fixes a hard limit to the satellite lifetime, limits the aperture of the telescope, and requires it to be launched cold, i.e. in a large vacuum vessel, thus consuming a major fraction of the mission mass budget and severely constraining the design. Radiative cooling allows the launch of telescopes which have no inherent limits to their lifetimes. However except at the shortest infrared wavelengths, pure radiative cooling is unlikely to achieve low enough temperatures to allow the efficient operation of infrared instruments and detectors, so that additional cooling methods must be employed, such as closed cycle mechanical systems (see, e.g., Orlowska, Bradshaw and Hieatt, 1990) or small amounts of on-board cryogens.

The degree to which radiative cooling is useful for a given mission obviously depends on the temperature to which the telescope optics can be cooled. Another important factor is the length of time after a warm launch which must elapse before the telescope achieves sufficiently low temperatures. In this paper we present simple guidelines for designing radiatively cooled space telescopes that must reach low temperatures, we explore the limits achieveable by such means, and we examine secondary issues such as the on-orbit cooling time. We also summarise results to date from a programme of more detailed model simulations. It appears that radiative cooling can indeed lower the temperatures of large-aperture telescopes sufficiently to make the technique of great interest for future IR space missions, while retaining advantages of size and duration, and can attain such low temperatures in quite reasonable times. We conclude by outlining possible missions of this type, the 2.5m EDISON proposal, the 1m Japanese SMIRT survey mission, and an intermediate-sized ESA/NASA mission.

2. Basic Principles of Optimisation: All-radiative Illustrations

In this section we use simple models to explore several basic design techniques for achieving low telescope temperatures by radiative cooling. Most of these techniques are routinely used in the design of cryogenically cooled satellites, for which heat input to the telescope and cryogens must be minimised to prolong the lifetime of the cryogen supply, but have not hitherto been seriously applied to possible cryogen-free missions. A few are very different from the approach needed for cryogenic missions. The models serve to illustrate the effects of these techniques, as a guide to their use for the design of radiatively cooled missions.

The models consider radiative effects only, and throughout this section of the paper we assume that individual components of the models (sunshades, tubes, shields) are each isothermal, i.e. have infinite internal conductivity.

2.1. "Basic Cylinder" Design

The simplest design for a satellite telescope in earth orbit is a cylinder, its bottom closed. Our initial model is covered on the outside with a layer of Teflon, which is silver coated on the underside. The Teflon is transparent, and the silver highly reflective, at visible wavelengths where the sun's emission peaks. This has the effect of making such a surface reject incoming solar-spectrum radiation, whether direct from the sun or reflected from the earth, leading to an absorption coefficient in the solar spectrum of ~ 0.08. However the Teflon has a high emissivity (~ 0.7) in the infrared, where the emission from a spacecraft at roughly normal temperatures must peak (see, e.g., McMordie, 1991, table 11-37).

The inner surface of the tube is painted black, giving it an emissivity as high or higher than the IR emissivity of the Teflon. We assume that the tube orbits with its open end pointing in such a way that no sunlight enters. For simplicity we ignore parasitic heating which may be caused by earth radiation entering the open tube, which for most orbits must happen at certain seasons of the year: for example, when the orbit passes through the earth-sun line. In a real satellite such heating input must be prevented by carefully designed aperture shields. As a further simplification we ignore the earth's radiative input to the outer surface, although this is highly significant in low orbits (see below, 2.6), and is a factor which cannot be ignored in any earth orbit. We let the cylindrical tube diameter be equal to its length, and for convenience set both equal to unity.

Figure 1(a) illustrates the factors affecting the balance between radiation absorbed and emitted by the structure and the resulting equilibrium temperature. The emitting surface is effectively the whole outer surface of the structure (including both ends of the tube, since the open end will have emissivity comparable to that of the closed end). For solar insolation of 1400 W m^{-2}, appropriate to earth orbit, such a tube will stabilise at 156 K. This is the sort of temperature which is conventionally expected for passively cooled designs.

2.2. Basic Cylinder with a Gold Surface: the Importance of Surface Properties

The great importance of the correct choice of surface properties is readily illustrated by changing those of the Basic Cylinder Model, e.g by simply replacing the outer layer with shiny vacuum-deposited gold. We now forfeit the high IR emissivity of the second-surface silvered Teflon, which increases the solar absorptivity by a factor of ~ 3. With IR emissivity $e \sim 0.03$ and solar absorptivity $a \sim 0.24$ (c.f., for example, Mason, 1988, Fig. 1) the only available radiator of significant efficiency is the black telescope tube, with the large outer surface now playing a minor role

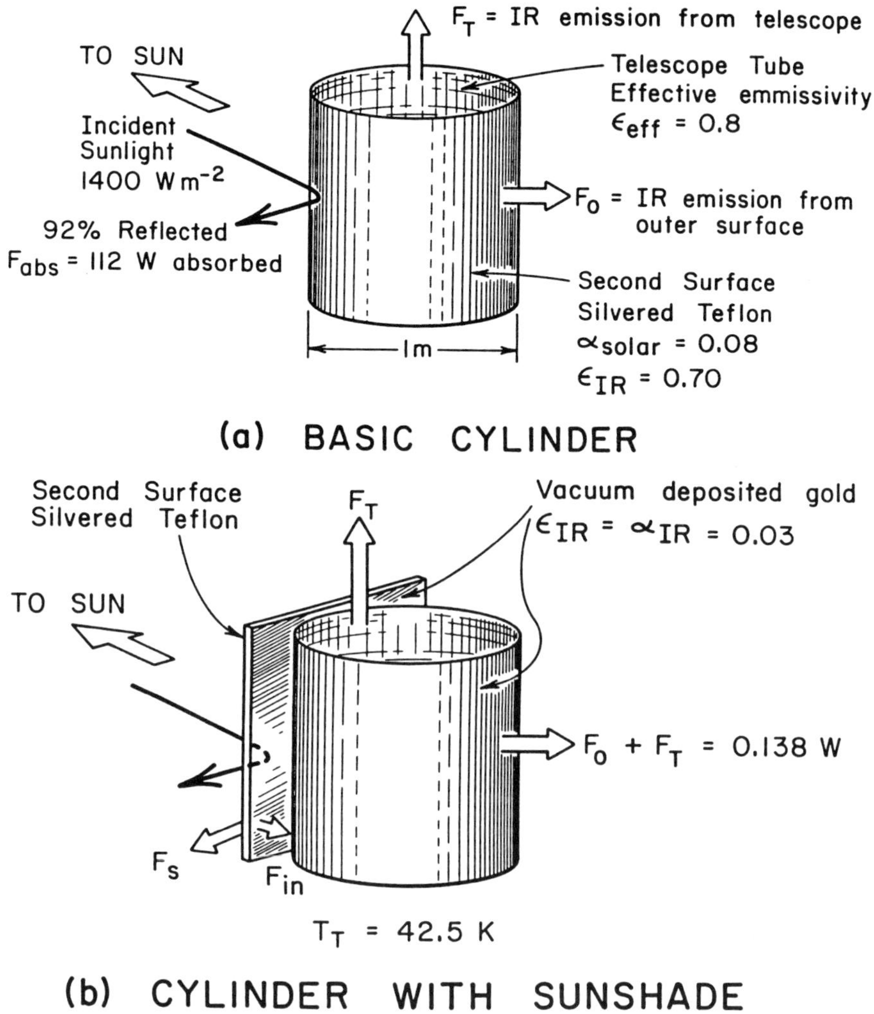

Fig. 1. Simple models of Radiatively Cooled telescopes. (a) Schematic illustration of the Basic Cylinder model with energy budget. (b) Schematic illustration of the Basic Cylinder with optimised sunshade; the energy budget is again indicated. See also Fig. 2.

in cooling. The resulting equilibrium temperature is ~ 300 K; all the advantage so far gained by going into space is therefore lost by the incorrect choice of surface material.

2.3. The Basic Cylinder with Sunshade

Much lower temperatures than that of the "Basic Cylinder" can readily be attained. A simple strategy is to take the vacuum-gold-covered variant of the design and convert its radiation environment from one dominated by (high-intensity) visible radiation to (low-intensity) IR radiation. This can be done by placing the satellite in the shadow of a suitably designed sunshade, as illustrated schematically in Figure 1(b).

This Figure again illustrates the factors determining the equilibrium temperature of the telescope tube, now only ~43 K. Principal amongst these are the efficiency of the sunshade at converting 1400 W m^{-2} of raw sunlight into a mere 4.6 W m^{-2} of IR emission in the anti-sunward direction, and the ease with which the shiny gold outer surface of the tube now rejects all but a couple of percent of that radiation.

Although we are tacitly ignoring a number of important factors which act to decrease the cooling which would be achieved in a real satellite, it is becoming evident that simple approaches offer large reductions in telescope temperature by radiative cooling alone.

2.4. Proportions of the Telescope Tube

The Basic Cylinder with Sunshade serves to illustrate the importance of the general proportions of the telescope. The incoming heat is absorbed on the outer surface of the cylindrical body, the area of which is proportional to its length; because the interior of the tube is black and the exterior is shiny, heat is lost primarily by radiation from the aperture, the area of which is proportional to the square of its diameter. Clearly, maximising the ratio of emission to absorption implies that radiatively cooled telescopes will normally be designed to be short and fat.

As in most areas there is a tradeoff here. The telescope tube is not completely black, but has for a bottom the shiny, low emissivity primary mirror. It is easy to see that the shorter the telescope tube, the lower will be the effective emissivity. However, in practise the enhancement of emissivity provided by the cavity of the tube more than makes up for this factor: for wall emissivities between 0.8 and 0.5 the effective emissivity of the tube calculated by Gouffe's method (e.g. LaRocca, 1978) is between 0.94 and 0.80. In what follows we adopt the latter as a conservative approximation.

2.5. Internal Radiation Shields

The next step is the insertion of internal radiation shields around the telescope tube. These are conventional in almost all cryostat design, and cryogenic space telescopes are no exception. Both the IRAS and ISO missions are protected by three such shields inside their vacuum vessels, enclosing the cryogen tanks. The shields are of high-strength, high-conductivity material, usually aluminium, with highly reflective surfaces – generally vacuum deposited gold, giving emissivities (and absorptivities) in the infrared of 0.03 and below.

In these missions cold gas from cryogen boiloff is used to cool the shields. This reduces the radiation emitted from their surfaces and hence the amount of heat transmitted in an inward direction. This cooling also carries away heat conducted along the straps which suspend the cryogen tanks, optical system and instruments within the cryostat. Since at this stage we are confining our attention to purely radiative effects, we omit consideration of the straps in this section.

In the cryogenic missions to date or proposed, Multi-Layer Insulation (MLI) is inserted between the shields. Our models omit this refinement and the thermal performance estimates are therefore fundamentally conservative. The omission of MLI is for several reasons, chief of which is uncertainty about the possibility of successfully degassing a large area of this material before cooling the optical system. In a mission which is launched cold, especially one constantly flushed by cryogen boiloff, gaseous contaminants from the MLI are largely frozen in or carried away. However these processes may not be as effective in a satellite which is launched warm and cooled on orbit, so that if outgassing continued into the cooling phase, condensation could present a serious hazard to the cold optical surfaces.

2.5.1. Cooling the Shields

Because our spacecraft is not a cryogenic system, the shields must be cooled by radiation. This must be achieved with minimal leakage of heat from radiators attached to the outer, warmer, shields to those connected to the inner, colder ones. This is most easily done by making the radiating surfaces coplanar. They must also be protected from radiation from outside sources. Since this is also true of the telescope tube (an important radiator in its own right) the simplest solution is to place the radiators around the telescope aperture, coplanar with the top of the telescope tube. One way of doing this is illustrated in Figure 2. The radiating surface Ar(i) is provided by the area of the annulus separating the (i)th cylindrical radiation shield from its inward neighbour. This annulus may be covered by a conductive metal plate attached to the shield but separated by a small gap from the neighbour, and blackened on its outward surface. As the annulus cools by radiation to space, it conducts heat away from the attached shield. High conductance along the shields is important for the efficiency of this process.

We note that if the annular plates are omitted the gaps between the shields, because of their small widths and great depths, are effectively quite black despite their highly reflective walls. Like the aperture leading into a black body cavity, these gaps can therefore act as efficient radiators of the emissions from their adjacent shields, cooling them in a manner which is, to first order, closely analogous to the effect of a material radiator covering the gap. (This is the configuration shown in Figure 2).

2.5.2. Radiation Transfer through the Shields

The shields constitute a series of large surfaces which are highly reflective, separated by gaps which are small relative to their extent. Radiative transfer between such surfaces is straightforwardly determined. Equation 7.72 of Bramson (1968), for example, reduces, for small values of the emissivities and absorptivities (assumed to be the same value, ϵ, for all the gold-covered shield surfaces) to

$$F(1-2) = A.\epsilon/2.\sigma.(T1^4 - T2^4) \tag{1}$$

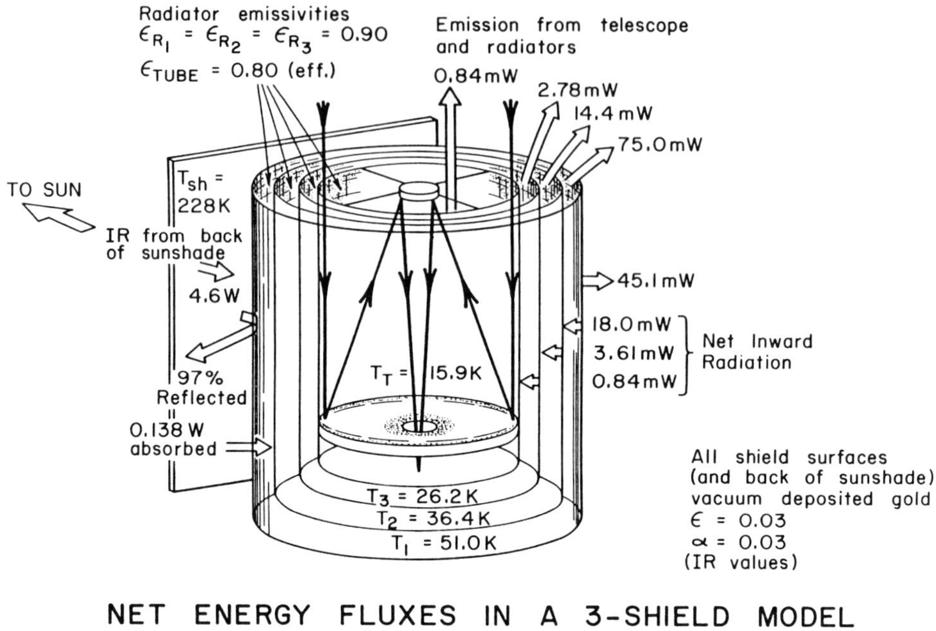

Fig. 2. Schematic illustration of a 3-shield satellite telescope with sunshade, showing main components of the energy budget.

where F(1–2) is the resultant of the bidirectional heat transfer between shields 1 and 2, A is the area (assumed the same for both shields), σ is the Stefan–Boltzmann constant and $T1$ and $T2$ are the absolute temperatures of the respective shields. The factor 1/2 comes from the fact that the photons emitted from either shield bounce back and forth many times between them and, in the limit of low and similar emissivities/absorptivities, have an even chance of being absorbed back into the surface which emitted them.

2.5.3. A Multi-Shield Model Telescope

Table I shows the results for multi-shield designs derived from the basic cylinder with sunshade shown in Figure 1(b), adding first the sunshade and then, one by one, the three internal radiation shields. In this simple model the individual shields are isothermal, i.e. assumed to have infinite internal conductivity. Figure 2 is a schematic representation of the 3-shield design, showing the resultant radiation fluxes.

Although the models are highly simplified, Table I indicates that by the radiative cooling technique low temperatures and large apertures can be achieved together. The 15.8K telescope is 0.61m in diameter, slightly larger than that on the ISO mission, but housed in a structure only 1.0m in diameter, half the size of the ISO cryostat. The model offers a gain of ~ 2 in the aperture which can be accommodated

TABLE I
Temperatures of illustrative all-radiative models (K)

Model	Outer Shield	Middle Shield	Inner Shield	Telescope Tube ($e = 0.8$)	Telescope Diameter (m)
Basic Cylinder				156 K	1.00
Basic Cylinder with Sunshade				42.5 K	1.00
BC + S′shade, 1 rad. shield	50.7 K			29.7 K	0.85
BC + S′shade, 2 rad. shields	51.0 K	36.2 K		21.6 K	0.71
BC + S′shade, 3 rad. shields	51.0 K	36.4 K	26.2 K	15.9 K	0.61

in a given outer cylinder. Further aperture can be gained at the expense of higher temperature by reducing the widths of the gaps between the shields, and hence the areas of the radiators. Since issues of source confusion imply that at longer wavelengths large apertures may provide greater sensitivity gains than do lower temperatures (section 1; see also Thronson *et al.*, 1992 in these proceedings) such a tradeoff may prove to be very desirable.

3. The Approach to a Realistic Radiatively-Cooled Mission

3.1. Complicating Factors

The simplifications of the illustrative models omit a number of important factors. A major one is the presence of the earth, which for a satellite in low orbit fills nearly half the sky with an efficient IR emitter at $\sim$ 240 K and, for half the orbit (unless it be a sun-synchronous polar one above the terminator) a diffuse source of solar-spectrum radiation composed of sunlight reflected from clouds and snow.

The importance of earth's influence can be appreciated by considering the on-orbit outer surface temperature of the Infrared Astronomical Satellite (IRAS), which is in a 900 km sun-synchronous circular orbit. Such an orbit keeps the satellite away from strong reflected sunlight. IRAS normally keeps its warm service module towards the earth and is further protected by an earth shield. Despite these precautions the surface temperature of the satellite averages 180 to 200 K (Mason, 1988). Largely for the very purpose of minimising earth radiation levels it is now planned that the SIRTF mission will be launched into a high (100,000 km) orbit. This is expected to reduce its surface temperature from the IRAS-like levels predicted when the mission was planned for LEO, to about 115 K, providing a major gain in mission duration (Lee, 1990). Even at 100,000 km altitude the earth still makes a significant contribution to the heat loads on a satellite. Its IR radiation flux density is ~0.8 W m^{-2}. The average intensity of reflected sunlight is slightly lower than this, but peaks strongly at full earth.

The surface temperature of SIRTF is nearly three times that derived in the previous section for our simple all-radiative models. Our omission of Earth radiation effects account for part of this difference, but most is caused by radiation and conduction from the warm service module, which must be maintained at near-room temperatures for the sake of its mechanisms, electronics and attitude control fuel. At 20 C = 293 K such a module, even if surfaced with vacuum-deposited gold, will emit about 12 W m^{-2} of IR radiation, a much higher intensity than that emitted by the back of the sunshade. This will double the total heat load absorbed by the outer surface of the satellite in Table I.

The surface temperature of the payload is strongly affected by conductive leaks from the service module, which must be strongly connected to the telescope assembly, to which it conveys the launch loads in all conventional designs. Similar leaks will occur down the support structure of the sunshade, if this is connected directly to the telescope assembly. The equilibrium temperature of the telescope itself is strongly affected by internal conductive heat leaks down the structures supporting it and the radiation shields and also by parasitic heating due to unwanted radiation entering the telescope tube and radiators (generally termed "aperture heating"). Other factors include conductive leaks down wiring to the instruments and active components, heat dissipation in these elements and heat leaks from the warm sections of mechanical refrigerators if these are used to cool the instruments and detectors.

The major effects, those of conductivity and of aperture heating, are discussed in the rest of this section.

3.2. Conductive Effects

The discussion so far has completely ignored the effects of conductivity in the structure of our hypothetical spacecraft. In practice there are two areas where conductivity is of major importance to the design. Because the spacecraft components must be attached to each other, to the bus and, during ascent, to the launch vehicle, substantial stuctural links must exist which join the subsystems being kept cold to others at much higher temperatures. The lower the conductivity of these links, the lower the temperature of the cooled systems (in cryogenic missions, the longer the cryogen lifetime). On the other hand, in radiatively cooled missions, effective cooling both of the radiation shields and of the primary mirror requires that HIGH conductivities should exist between these components and their radiators.

3.2.1. Low-Conductivity Links to Supporting Structures, and a Scaling Law

A good deal of work has been carried out in this area, in the development both of scientific missions and of liquid-gas storage systems for use on long-duration missions such as Space Station Freedom. Most or all current designs make use of tensioned straps suspending the mass to be kept thermally isolated from an outer structure, usually the cryostat jacket, since this must be stoutly made and so has considerable mechanical strength available for support. Such straps, of glass-fibre (GRP), alone or in combination with carbon-fibre (CRP) or, as recently proposed, of alumina-based composite, have very high tensile strengths and low conductivities. Fabricated as loops or as chains (e.g. Emery and Burgio, 1982, p72) they can therefore act as effective isolators for the cold structure. As a strap passes through each radiation shield, part of the heat being conducted down it is absorbed by the cooled shield, to which the strap is thermally well-connected.

The design of support straps is generally driven by the need to keep the lowest lateral and (especially) axial vibration frequencies of the suspended mass above some critical value set by the launcher specification. For a shared launch on an Ariane and for launch on a Delta vehicle this axial frequency is 31 - 35 Hz. The required cross section area of a strap of given length is proportional to the square of this frequency, so that major reductions in conductivity can be achieved if a more relaxed vibration specification can be accepted. Such a relaxation is offered by the specification of the Titan IV launcher proposed for SIRTF (minimum axial frequency $\sim$ 10 Hz), although some of the benefit thereof is lost because the strap design becomes driven by the avoidance of fatigue problems (Lee, 1990).

The straps generate an important scaling law for radiatively cooled missions. Both the area through which radiative leaks occur, and the radiating area available for removing heat, scale as the square of the telescope dimensions. The ratio of heat input and output by radiation is therefore constant, i.e. purely radiative effects on the temperature are scale invariant. On the other hand, however, the strength required of a strap is proportional to the suspended mass, and if all other factors are equal, this mass will vary as the cube of the telescope diameter. The strength of a strap, and its conductivity, are both proportional to its cross section area, and so the conductive component of the heat leak into the cold structure will also scale as the cube of the diameter. Scaling up the size of the structure will therefore produce an increase in the fraction of the heat leak which is conductive, and the equilibrium temperature will accordingly rise slowly with the diameter of the telescope.

3.2.2. Conductive Connections to Radiators: Materials and Mechanisms

The design objective here must be to have the highest possible conductivity between the component being cooled and the radiator doing the cooling, so that their temperature difference is minimised. The reason for this is that the heat flow to the radiator is set by the temperature gradient down the conductive link, which is present because the cooled component remains warmer than the radiator. Minimising this temperature difference results in a lower temperature for the cooled component relative to its radiator.

A simple approach to this problem is to fabricate the shields and the telescope tube out of a physically stiff material with a high ratio of conductivity to density. The best such materials known to us are low-impurity aluminium alloys such as 6061-T6 or 6063-T5, which also offer a useful increase in conductivity below about 100 K (c.f. Hust, 1983, Fig. 4-27 for examples). Aluminium tubes and shields have been assumed in all models studied in our work to date. An interesting alternative, which we have not yet explored, is the use of heat pipes, employing wicking fluids chosen to be suitable for the particular temperature range being considered. Substantial increases in conductivity with minimal increases in mass may be achievable thereby (e.g. Williamson, 1987).

High conductivity to the radiators is another reason why a short fat telescope tube (section 2.4) is advantageous in a radiatively cooled satellite. The conductivity of the telescope tube is inversely proportional to its length, so the shorter the tube the lower the temperature excess of the primary mirror over the upper tube at equilibrium. As is shown below, in section 4, minimising the conductivity of the telescope tube is also important for reducing the on-orbit cooling time required for the telescope to reach working temperatures.

Note that this is an area where the design principles for an optimised radiatively cooled telescope differ sharply from those appropriate to cryogenically-cooled missions. The latter are designed to have low conductance down the telescope tube, because the top of the tube is subject to absorbing radiant energy from sources both local (e.g. the aperture shield) and external (Moon, Jupiter, scattered earthlight) and conducted heat, e.g. from the cryostat surface. The telescope tubes on cryogenically-cooled satellites therefore include a ring of insulating material some way down the tube, or baffle, to prevent heat flow down the tube from adding to the cryogen boiloff rate and thereby shortening the mission lifetime. The effect of such a baffle splitter is illustrated in the ISO phase A design, in the original, two-cryogen version, where it maintained the upper telescope tube at 30 K, the lower tube at 17 K (Emery and Burgio, 1982; p 75).

3.2.3. Conductivity of the Main Support Structure

The telescope and its radiation shields must be supported during launch and stiffly connected to its service module throughout the mission. Thus a strong supporting structure is required.

In cryogenic missions much of this strength is supplied by the cryostat vacuum jacket, and further design becomes a matter of providing a rigid low-conductivity connection between this jacket and the service module or bus. A warm-launched radiatively cooled telescope does not require this stout, heavy vacuum jacket and the design of the support structure can be optimised for structural stiffness and low conduction only. It is therefore likely that, with careful design, major mass savings can be made and somewhat lower temperatures can be achieved, for example, for the outer surface and for the outer attachment points of the supporting straps, in a radiatively-cooled mission than in its cryogenic counterparts.

3.3. Aperture Shields and Parasitic Heating

Telescopes in space must be protected from direct and scattered sunlight, while those in Earth orbit must also be protected from Earthlight entering the telescope tube. Not only do these sources generate scattered background light, but in a cooled mission they can generate a very significant heating effect. Such telescopes must also be protected from the relatively powerful infrared emission from the back of the sunshade.

Cold telescopes in Earth orbit must therefore be fitted with an efficient Aperture Earth Shield ("AES" in our notation). Aspects of the design of such a shield are discussed by Mason (1988) for the case of IRAS and are illustrated in Figure 3. The shield is an inverted truncated cone opening outwards and cut off at an angle which determines the minimum sun avoidance angle. In order to protect the outside of the AES from direct sunlight and the telescope aperture from radiation from the back of the sunshade, the latter should extend very close to or up to the cutoff plane of the AES but not beyond it.

The angle of opening of the cone together with its height ensures that Earthlight entering it is reflected directly away into space. The satellite orbit determines the combination of sun- and Earth-avoidance angles for which the AES must be designed. The larger the semi-angle of the Earth above the orbital plane (i.e. the closer the satellite comes to the Earth) the larger must be the size of the AES in order to exclude sunlight and reject Earthlight.

The inner surface of the AES is directly visible from the telescope aperture (and radiators) and must therefore also be cooled to reduce the radiation which they absorb from it. In the case of IRAS the AES had its own radiators which maintain the inside conical surface at $\sim$ 95K (Mason, 1988). The smaller the area of the cone visible from the telescope tube, the smaller the parasitic heating by its emissions. Because the size of the AES is reduced as the minimum orbital altitude is increased, reduction of this parasitic heating is another important advantage of staying well away from Earth at all times.

As indicated above, a properly designed AES for a radiatively cooled telescope must protect not just the telescope itself, but also the annular radiators on the radiation shields, from heating by the Earth, sun and from the back of the sunshade. Such an aperture shield on a radiatively cooled mission must therefore be markedly larger than one appropriate to a cryogenically-cooled mission with the same overall diameter in the same orbit, as in the cryogenic case only the (smaller) telescope aperture needs to be protected. The bulkiness of the AES on a radiatively-cooled mission is likely to be the overall limiting factor in fitting the largest possible telescope into a given launcher fairing. This again is a powerful driver for radiatively cooled missions, in particular, to be placed in the highest possible orbits.

This encourages the consideration in mission planning of the use of non-Earth orbits. For example, in an L2 halo orbit, Earth and sun remain within 26 degrees of

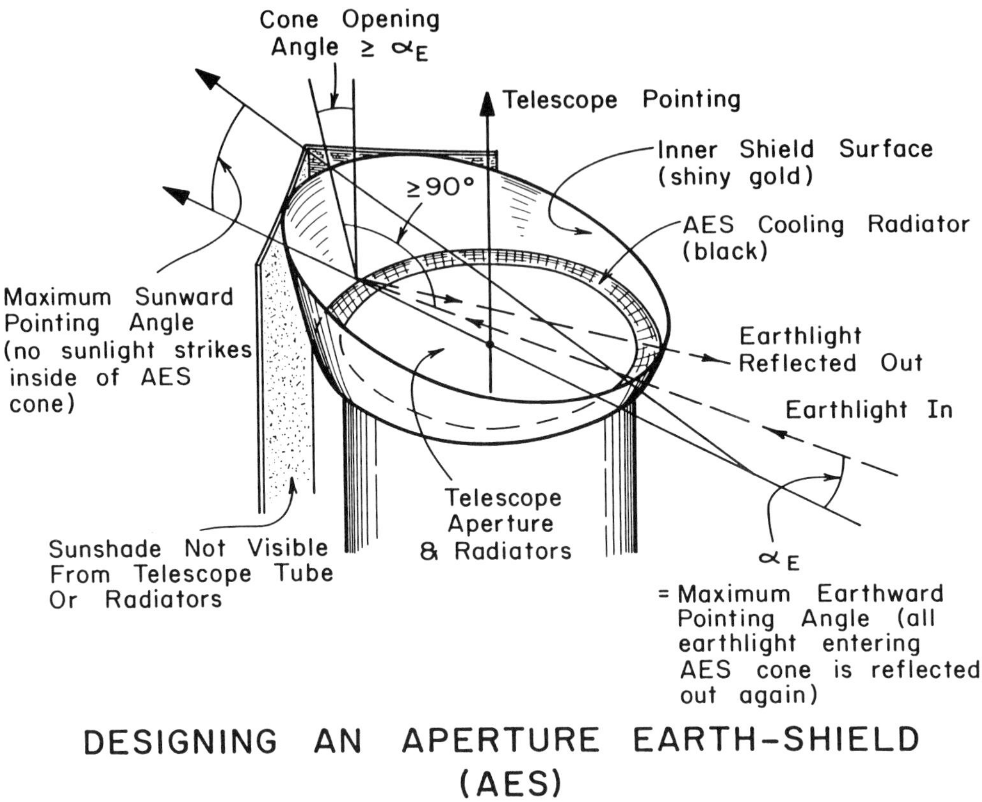

Fig. 3. Illustrative outline of the design principles of an Aperture Earth-Shield (AES).

one another on the sky. If sun-avoidance constraints are applied relative to Earth as well, all that is required is that the top of the telescope tube and the radiators be cut off obliquely, at an angle slightly greater than the minimum sun/earth avoidance angle. An AES is not strictly necessary at all, which yields important savings in launcher volume, parasitic heating, complexity of design and of mass.

4. On-Orbit Cooling Times

4.1. Simple General Model and Basic Equations

A simple but reasonably realistic model of the cooling process for a satellite telescope is presented in Figure 4. It comprises just a radiator, here represented as a nearly-black disc with the same diameter as the aperture of the telescope tube, joined by a conductive link – the tube walls – to the mass being cooled, which to a good approximation is the primary mirror, its support structure and the instrument package.

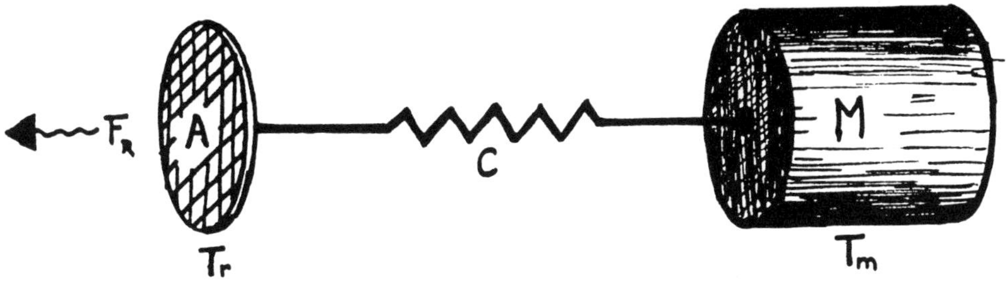

Fig. 4. The simple thermal model used in the cooling-curve calculations. It comprises a radiator, area A, representing the telescope aperture, a conductive connection C, which in practice would be provided by the telescope tube walls, and the mass M to be cooled, which in a real telescope would comprise the primary mirror, its support structure and the instruments.

In general:

$$dT/dt = -(cooling\ power)/(thermal\ capacity)$$

$$dT/dt = -P_{cool}/(mC_p) \tag{2}$$

where m is the mass, C_p the specific heat, of the mass being cooled.

The power conducted down a tube of conductance G is

$$P_{cool} = (T_m - T_r)G \tag{3}$$

where T_m is the temperature of the mass being cooled and T_r that of the radiator.

The cooling power of the latter is

$$P_r = A\epsilon\sigma T_r^4 \tag{4}$$

where A is the radiator area, ϵ the effective emissivity and σ the Stefan-Boltzmann constant.

4.2. Illustrative Initial Cooling Rates

4.2.1. Simplest Case: Infinite Tube Conductivity

The maximum possible cooling rate is achieved if the conductivity k is assumed infinite, when $T_m = T_r$ and $P_{cool} = P_r$, so that cooling of the telescope is radiatively limited.

Then from (1) and (3) in 4.1:

$$-dT/dt = (A\epsilon\sigma T_r^4)/(mC_p)$$

For an example consider a 2.5 m telescope, with $A = 4.9\ \mathrm{m}^2$, $\epsilon = 0.9$ (see below, section 4.2.2), $T_r = 300$ K. Let $m = 2500$ kg, $C_p = 1000\ \mathrm{J\ kg^{-1}\ K^{-1}}$ (approximately appropriate for aluminium).

Then

$$-dT/dt = 8.1\ 10^{-4}\text{Ks}^{-1} = 2.9\text{Kh}^{-1}$$

This initial cooling rate is well illustrated by the time-to-zero,

$$t(0) = -T/(dT/dt)$$

For this model $t(0) \sim 100$ hours. Although this quantity is only distantly related to the cooling time of a real telescope, it is already evident that such times are likely to be short.

4.2.2. Cooling With a Realistic Telescope Tube

We may now take account of the finite conductivity of the telescope tube in an approximate way, as follows.

Assume a tube wall thickness of 2 mm, diameter 2.5 m, length 4 m. The conductance is

$$G = CA/L$$

where C is the specific conductivity, A the cross-section area, L the length. For the model, assuming the tube is made of 6063-T5 aluminium with C = 215 W m^{-1} K^{-1}, we have $G = 0.844$ W K^{-1}. For a tube of these proportions the effective emissivity of the aperture, estimated by Gouffe's method (cf. LaRocca, 1978) and assuming zero emissivity for the base of the tube (the telescope mirror) ranges from 0.97 to 0.90 for wall emissivities of 0.8 to 0.5. We adopt the lower value of the effective emissivity as a conservative approximation.

Assuming that T_m changes very much more slowly than T_r (since only the small thermal inertia of the thin-walled tube limits dT_r/dt) a slowly changing equilibrium will arise soon after launch, giving, from (2) and (3) in 4.1:

$$A\epsilon\sigma T_r^4 = G(T_m - T_r)$$

With $T_m \sim 300$ K this may be solved for T_r, giving $T_r \sim 150$ K, whence $P_{cool} = P_r \sim 130$ W. So

$$\begin{aligned} dT_m/dt &= P_{cool}/(mC_p) \\ &= 130/(2.5\ 10^6)\text{Ks}^{-1} \\ &\sim 0.2\text{Kh}^{-1}. \end{aligned}$$

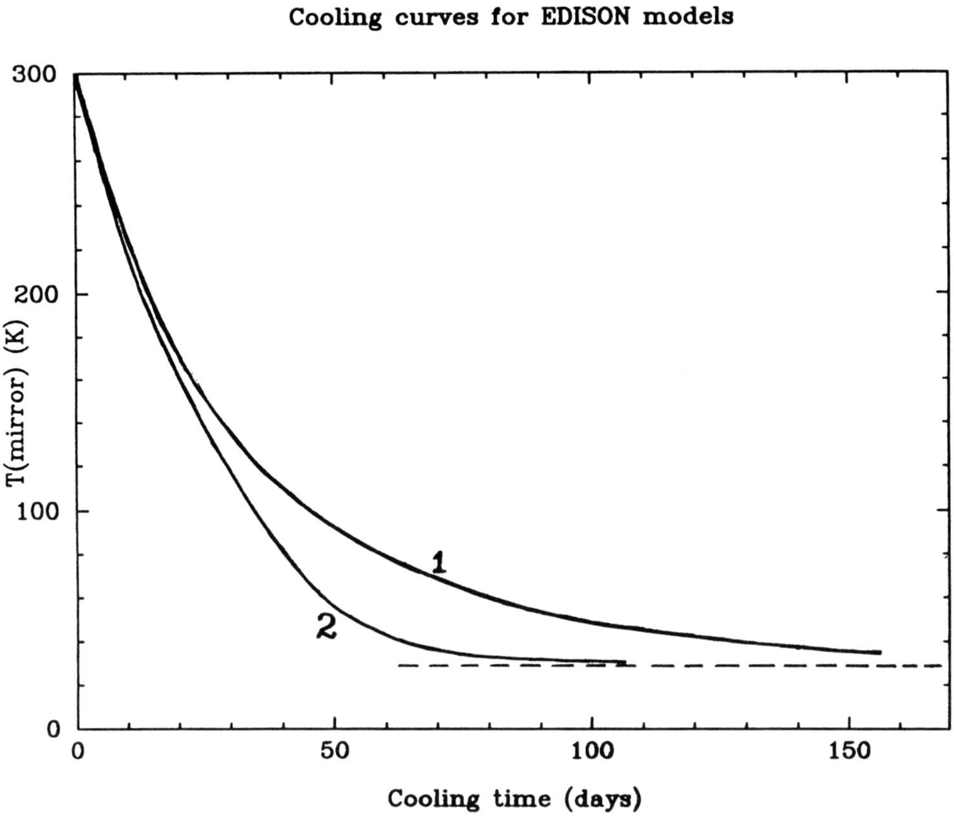

Fig. 5. Representative cooling curves for models for the 2.5 m EDISON mission. Curve 1 is a conservative model, comprising a 900 kg quartz primary mirror, 600 kg of carbon-fibre composite support structure, and 300 kg of aluminium instruments (6061-T6 alloy). Curve 2 is for a model optimised for rapid cooling, with a 600 kg beryllium mirror, 400 kg of carbon-composite support structure and 150 kg of aluminium instruments. The curves take into account the variations with temperature of the conductivity of the aluminium telescope tube (assumed to be 6061-T6 alloy) and of the specific heats of the quartz, carbon-composite, aluminium and beryllium components.

4.3. More Detailed Cooling Curves for Space Telescopes

In any attempt to predict the cooling curves of real missions it is essential to take into account the strong variations in both conductivity and specific heat which occur in most materials at temperatures below about 150 K. Using conductivity data from Hust (1983) and specific heat (C_p) data from Sparks (1983) we have derived detailed cooling curves for several possible satellite models. Two of these are illustrated in Figure 5. Model 1 is a conservative design (c.f. 6.2 for more details) comprising a 900 kg quartz primary mirror supported by and connected to 600 kg of carbon-fibre backing structure, and 300 kg of aluminium instruments. The predicted cooling time to 70 K is $\sim$ 80 days and the model is within a few degrees of its equilibrium temperature by 200 days.

Such times are lengthy by comparison with the lifetimes of cryogenic satellites launched so far, but appear by no means excessive in a mission with a lifetime of several years, as would be appropriate for a radiatively cooled telescope. However, if shorter cooling times must be achieved, choice of material and careful design offer opportunities to improve matters. To illustrate this Model 2 comprises a 600 kg beryllium mirror supported on 400 kg of CRP backing structure with 150 kg of aluminium instruments. Because the specific heat of beryllium drops dramatically at temperatures below $\sim$ 100 K the cooling time for this model is much shorter: as the curve shows, it takes less than 50 days to reach 70 K and is close to its final temperature in $\sim$ 100 days. Overall, it is clear that any sensibly designed radiatively cooled satellite telescope should have no difficulty in coming close to its equilibrium temperature in much less than a year after launch.

5. Studies for Real Missions

5.1. POIROT: The Proposal for ESA'S Second Medium-Sized Mission

Early investigations into the potential of passively cooled IR missions led to a proposal to the ESA for its second medium-sized mission M2 in November 1989. This proposal, by the Royal Observatory, Edinburgh, was for a 1.5 m radiatively- cooled mission named POIROT (**P**assively cooled **O**rbital **IR** **O**bservatory **T**elescope). The proposed mission was strictly constrained to the resource guidelines of the ESA mission: it had to share a dual launch on an Ariane 4 and would need to coexist with the low-precision pointing of the only available bus, that under development for the ESA cornerstone, the X-ray Medium Mission (XMM).

As proposed, the 1.5 m telescope was to have used the same 24 h elliptical orbit planned for ISO. The 900 kg payload was to have included a fast lightweight telescope with a complement of instruments, two imagers and two spectrometers, all with detectors using silicon array technology and cooled by two- and three-stage Stirling/Stirling/Joule-Kelvin refrigerators under development at the Rutherford Appleton Laboratory and currently space qualified in single-stage form. The wavelength coverage was thus confined to operations shortwards of $\sim$ 32 microns, although possible extension to longer wavelengths was contemplated. Fine guiding and pointing was intended to be applied through control of the secondary mirror, at least at short wavelengths. A schematic outline of this mission has been published elsewhere (Davis, Hawarden and Mountain, 1991). Thermal analysis of a simple model using spreadsheets on a personal computer indicated an equilibrium temperature of $\sim$ 37 K, appropriate to the proposed wavelength coverage.

While well received by ESA's Astronomy Working Group, the proposal was not finally selected for further study by the Agency. Nevertheless we continued to analyse the design after the ESA process was concuded, as it was far from fully developed. At this stage it became apparent that the original proposal was flawed in two ways. The optimisation of the radiators cooling the radiation shields had been incomplete, and a practical design would not have been able to achieve the effective radiative areas assumed in the analysis without serious radiation leaks from warm to cold areas. The second, related, difficulty would have arisen from the the orbit proposed, which would have caused large transient heat inputs to the radiators at the low perigee passage. Even if an optimally designed aperture shield were fitted, serious aperture heating would be unavoidable around perigee, while such a shield would still have violated the space constraints of the launcher unless the aperture of the telescope was drastically reduced. Alternative orbits (with higher perigees) would involve substantial mass reductions and, again, would make the telescope so small that the performance of the mission would be seriously compromised.

Nonetheless we note that these problems could have been eliminated for a rather small increase in the mission budget. This is discussed in section 8.2 along with the options opened by the possibility of an ESA - NASA collaboration on a mission on this scale.

5.2. A Radiatively Cooled IR Telescope in High Earth Orbit

In 1990 a low-level thermal study was commenced at the Marshall Space Flight Center (MSFC) as part of a broader programme to understand the properties of telescopes in High Earth Orbit and beyond. From this study we have gained insights into the optimisation of numerous design factors for a passively cooled telescope by using a detailed and rigorous thermal analysis. The analysers used were the Thermal Radiation Analysis System (TRASYS) (JSC version P22) and the System Improved Numerical Differencing Analyser (SINDA) (Gaski version of 1987), installed on the Cray X-MP computer at MSFC.

The model, the Radiatively Cooled IR Telescope (RCIRT), has evolved to a considerable degree but is as yet neither fully realistic nor fully optimised. The results so far, however, have enabled us to identify the changes needed for this optimisation, while greater realism is mainly a matter of attachment of the additional components which the real world demands, such as a warm service module (some of the effects of which are already incorporated).

5.2.1. Details of the RCIRT Model

The RCIRT is a three-shield design for a 2.5 m aperture telescope. The structure is protected by a large flat sunshade, designed as outlined in 2.3. The payload itself, constructed entirely of 6061-T6 aluminium alloy, comprises the three shields and the central telescope tube, which contains a primary mirror of aluminium plates separated by a honeycomb structure. Behind the primary mirror is an instrument bay enclosed within an extension of the telescope tube; 50 mW of heating is applied to the rear face of the primary mirror to simulate the effects of parasitic heating from the instruments.

The model is very similar to the illustration in Figure 2, scaled up for a 2.5m primary mirror diameter. However it was originally run in an elliptical 24h orbit, and because adding a realistic aperture shield was considered to be too complicated for an initial experiment of this sort, the lengths of the radiation shields decrease by approximately 60 cm per shield from outboard to inboard, to prevent Earthshine from entering the radiators and telescope tube during the low perigee passes. The tube is thereby also protected from IR radiation from the back of the sunshade. The resulting structure is 3.5m in diameter and 6.3m long (excluding the sunshade, which is 7.6m long). The radiation shields are currently cooled by emission from their separating gaps, which are 16.7 cm wide. This cooling effect was actually discovered as a result of our analysis of the results from these initial models and has been discussed earlier, in section 2.5.1. The staggered structure of the radiation shields carries an appreciable penalty, as the inner shields and telescope tube can "see" the inside surfaces of the outer shields and do, in fact, absorb a very significant amount of radiation from them. The inner shields and tube are attached to the outermost shield by 24 straps of S2 fibreglass, sized for an assumed launch on a Titan IV (see section 3.2.1) of a suspended mass of 2800 kg (see section 6), giving total cross-section 20.4 cm^2. There is at present no outer support structure, so the important conductive leak from the service module via such structure is currently absent. However the IR radiation that would be absorbed from a warm service module is simulated by an applied heat load of 4 W distributed over the aft outer surface of the payload module.

The material thicknesses in the model are currently several times larger than would be the case in a real payload. This means that longitudinal conductivities are unrealistically high, producing slightly optimistic temperatures and tending to shorten cooling times, but the total mass is also markedly too high, which lengthens the latter times by a more than compensatory amount.

After the initial runs assuming an elliptical 24 h orbit, a 100,000 km radius equatorial circular orbit (similar to that currently planned for SIRTF) was adopted. This transition involved some changes to the simulation system, which was designed for low orbits in the first instance. The orbit currently in use is a springtime "worst case", with the earth-sun line in the orbital plane. Three-sigma high values have been used for terrestrial and solar radiation inputs.

Several modifications to the models were made in the light of analyses of results from the initial simulation runs, for which several small tools were developed, including a Monte-Carlo simulation of photon flux through areas at various positions relative to cylindrical emitting surfaces. The satellite surface properties are currently semi-optimised, with second-surface silvered Teflon on the sunward face of the sunshade, gold multilayer insulation on the outer surface of the tube assembly and gold tape on all other surfaces. Changes included blackening the inside of the telescope tube after verifying that the resulting absorbed radiation from the inside surfaces of the outer radiation shields was more than compensated for by the increased emissivity.

5.2.2. Current Results

Figure 6 shows the evolution of the temperature of the outer surface of the RCIRT over 50 days after launch. The initial rapid cooldown is followed by stability under

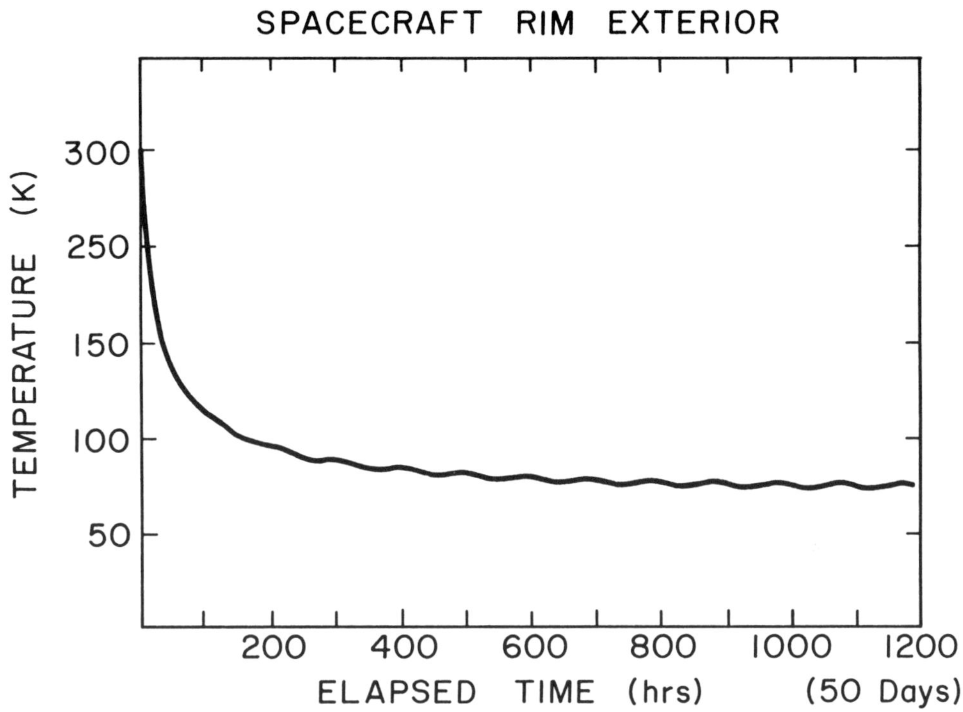

Fig. 6. The temperature of the outer surface of the RCIRT model as simulated at Marshall Space Flight Center, showing the first 50 days in orbit. Note the small fluctuations caused by changes in earth illumination around the orbit.

small-amplitude fluctuations caused by changes in Earth-irradiation between full and new earth, plus the (unresolved) effects of the eclipse at new earth.

The mean equilibrium temperature of the outer surface is $\sim$ 75 K, which is lower than would be the case in a real satellite; the expected temperature for SIRTF is $\sim$ 115 K. The difference is at least in part due to our omission of conduction from the service module and sunshade.

Figure 7 is the cooling curve for the open end of the telescope tube. This is still cooling after 50 days – as would be expected from the cooling curves in Section 4 – and in fact the temperature is still dropping after 100 days, at which stage it is below 38 K. Because the telescope tube is more than four times more massive than would be the case in a realistic design, the temperature drop in this figure is probably too slow.

Figure 8 shows the cooling of the primary mirror, in a somewhat artificial configuration with no conductive (i.e. purely radiative) connection to the telescope tube. This is very much a worst-case arrangement, but serves to indicate the adverse effects of a relatively low-conductivity mirror support structure. After 100 days on orbit the primary mirror is at about 67 K and is still cooling steadily. Again, this is of course in good general agreement with the predictions of the cooling-curve

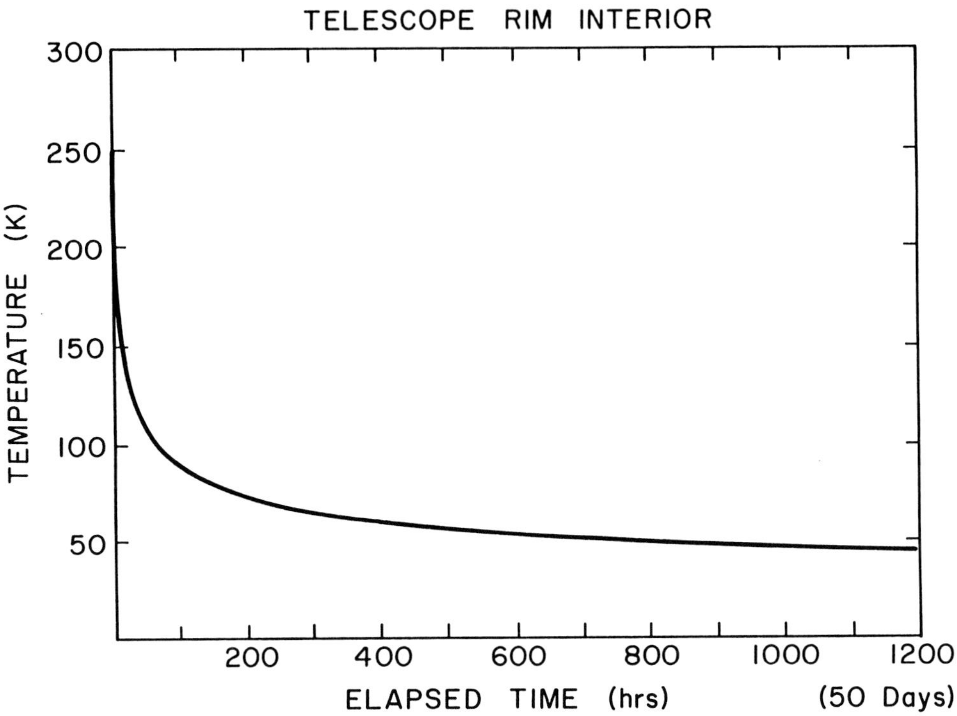

Fig. 7. The temperature of the open end of the telescope in the RCIRT model during its first 50 days in orbit. By 100 days the temperature of the tube has dropped below 38 K. This Figure illustrates the behaviour of the primary cooling radiator for the telescope assembly.

models in Section 4.

We do not yet have a definitive answer to the final equilibrium temperature of the optics in this model. The temperature of the upper telescope tube is a lower limit at any time in the cooling process, and we can infer from the likely heat loads and the tube conductivity that the equilibrium temperature of the primary mirror with good conductive connection to a tube of realistic thickness would probably be a little below the 38 K reached by the tube in the present configuration.

5.2.3. *Planned Developments*

By the time this paper appears in press several important enhancements to the model will have been implemented. Sizing the radiation shields so that their tops lie in the same plane, and adding an aperture shield to remove the current small heating load from the earth, and the larger one from the back of the sunshade, will reduce the heat input to the telescope tube by more than a factor of two. The current (rough) estimate of the equilibrium temperature after this change is 29 K. Other changes will lose some of this gain. From the SIRTF predictions a realistic simulation of an outer support structure connecting the payload to the

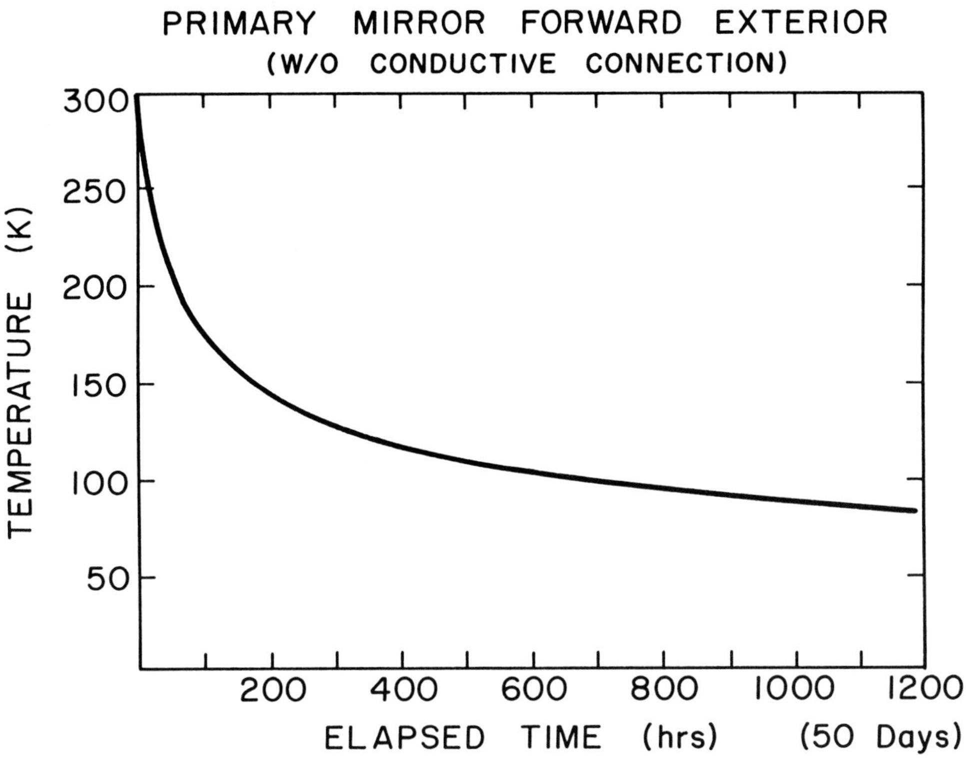

Fig. 8. The cooling curve for the primary mirror of the RCIRT model during the first 50 days in orbit. For this curve the mirror had no conductive connection to the telescope tube, but cooled by radiation to the tube walls only. This worst case simulation nevertheless reaches 67 K in 100 days, at which time it is still cooling quite steadily. In a real spacecraft the mirror would be more closely connected to the telescope tube, its radiator.

service module may be expected to raise the surface temperature from the present 75 K to $\sim$ 110 K, although lower temperatures ought to be be possible with careful optimisation of the structure (see section 3.3). The higher outer surface temperature could raise the equilibrium primary mirror temperature to as high as 42 K, although lower values are probable.

The MSFC models therefore imply that earlier, simpler models had indeed yielded reasonable approximations to the sort of temperatures which could be obtained by pure radiative cooling, given optimisation in several areas. They also support the simple models of the cooling process presented in Section 4, confirming that missions such as these can probably achieve working temperatures in times of from tens to hundreds of days.

6. The EDISON Design Proposal

The MSFC programme is directed to understanding the dynamic thermal properties of telescopes in High Earth Orbits, for applications to various possible NASA missions. The RCIRT study is also applicable to the EDISON design proposal, which currently comprises a 3 m class telescope in HEO. This proposal now contains a reasonable amount of detail and provides an overview of the likely properties of a large mission of this type. It should be noted, however, that it is a markedly conservative design, and currently does NOT exploit several further possibilities for improving its performance. Thus, the design contains no heat pipes; it is for a satellite in the "conventional" 100,000 km HEO, and so misses that advantages which ought to be gained from moving to a similar energy orbit around the solar Lagrangian point (see 7, below). Several aspects of the proposal are discussed elsewhere in these proceedings (Thronson *et al.*, 1993); here we deal with the technical issues only.

6.1. Outline Description of the EDISON Proposal

EDISON, an artist's impression of which is shown in Figure 9, closely resembles the configuration which we believe the RCIRT model will have assumed at the end of the current cycle of optimisation and development towards realism. We outline some of the main differences.

6.1.1. Main Structure

Like the RCIRT it is a conventional (earth-orbit) design with three shields and telescope tube; these are of 2 mm thick 6063-T5 aluminium alloy. The tube and shields are flush-topped and cooled by "gap" radiators. The upper end is fitted with an Aperture Earth-Shield (AES), radiatively cooled to ~ 80 K, which protects both the telescope and the radiators from earth-heating. Because of the high orbital altitude, this shield is quite small (see section 3.4). The telescope mirrors and the 100 kg instrument bay are supported on a common support structure of CRP; the tube is attached to the inside and a thin layer of gold-covered MLI surrounds the whole on the outside. This assembly is necessary for structural strength, but because it effectively inserts another, partial, layer of thermal protection it would give significantly reduced telescope temperatures relative to the simple structure of the RCIRT model.

The inner structure and the radiation shields are suspended on straps of S2 fibreglass, sized for a 2800 kg suspended mass and for launch on a Titan IV- Centaur vehicle. At their outer ends they are attached to a lattice supporting structure (the Outer Support Structure) which is in turn enwrapped in MLI.

6.1.2. Self-Cooled Secondary Mirror

At the upper end of the tube, fitted with a protective baffle around its outboard surface, is the on-axis $f/15$ Cassegrain secondary mirror. The baffle, coated with

Fig. 9. Artist's impression of the telescope proposed for the EDISON mission in a high Earth orbit (HEO).

vacuum-gold, is designed to prevent radiation from the AES from impinging on the back of the secondary. The support system for this mirror would be designed for minimum conductance between it and the telescope tube, to which the supports would be thermally contacted. Using thin tensile structures, extremely low conductances should be obtainable. The back (outward) surface of the secondary assembly is black, so that the whole is cooled by its own radiator to space. For a radiation-only design, assuming 30% of the total area is black and faces space, the rest shiny gold and facing the telescope tube and the AES, an equilibrium temperature of $\sim$ 17 K is achieved with the upper telescope tube at $\sim$ 37 K, though a number in the low twenties kelvin is more likely. If the secondary can indeed achieve this degree of cooling it would be effectively removed as a source of background emission relative to the contribution from the primary, thereby halving the effective emissivity of the telescope optics as seen by the detectors.

6.2. Mass Budget

The current mass budget for EDISON assumes a 2.5 m aperture $f/1.2$ primary mirror. It is derived almost entirely by scaling from existing designs for cryogenic missions. It is therefore likely to be roughly realistic, if somewhat conservative.

This mass budget was used to determine the strap thicknesses for the RCIRT model.

The major source of baseline information was the ISO Phase A Study. Cross checks and some basic design information were obtained from papers describing SIRTF, in particular Lee (1990). The derivation of each quantity in the Mass Budget is briefly outlined.

6.2.1. Suspended Mass

1. COMPONENTS FULLY COOLED BY RADIATION FROM THE TELESCOPE TUBE:
   - Primary Mirror: 900 kg.
     The 2.4 m "honeycomb" quartz primary mirror of the Hubble Space Telescope has a mass of 818 kg. The primary mirror for the US-German airborne telescope proposal SOFIA is currently planned at 800 - 900 kg for an identical optical design and similar optical quality to that proposed for EDISON.
   - Mirror support system: 400 kg.
     The optical support system in ISO (phase A) is 21 kg for a telescope mass of 56 kg. Scaling to our primary mirror mass gives $\sim$ 340 kg. Credibility is lent to this exercise by the mass of the mirror support system in SOFIA, which must cope with accelerations in varying directions, and totals $\sim$ 500 kg.
   - Telescope tube: 200 kg.
     This is a tube of 2 mm thick 6063-T5 aluminium, 4 m long and closed at one end. Additional conductivity can be added at a cost of $\sim$ 80 kg per mm of thickness.
   - Scientific Instruments: 100 kg.
     The instrument mass on ISO is $\sim$ 30 kg. 100 kg is more appropriate for a long-duration mission which ought to be more versatile and durable.
   - Common Support Structure: 200 kg.
     This is the CRP framework which would support the telescope optics, tube and instruments and be attached to the outside world by the support straps. It differs from the other telescope-cooled components in that it could have a much better thermal connection to the radiator than the rest. On ISO, which has a suspended mass of 700 kg, the CSS masses 46 kg. Scaling by suspended mass then gives $\sim$ 170 kg for EDISON.

   TOTAL MASS COOLED BY THE TELESCOPE TUBE: 1800 kg.

2. COMPONENTS INDEPENDENTLY COOLED:
   - Secondary mirror assembly: 100 kg.
     The mass is a rough estimate, probably too high.
   - Radiation shields: 900 kg.
     This is based on our currently favoured design, with the tops of the shields flush with the top of the telescope tube. The three shields are of 2 mm 6063-

T5 aluminium. They are not cooled by the telescope tube but by their own radiators (currently just the gaps between each shield and the next).

TOTAL SUSPENDED MASS: 2800 kg.

6.2.2. Masses Not Suspended on Straps

- Outer structure: 600 kg.
  Scaling the outer vacuum jacket of the ISO cryostat by the suspended mass implies a structural mass of $\sim$ 1400 kg. Substantial reductions can be achieved when the only goals in the design of the structure are physical support and low conductivity, and when advanced composite materials can be used; hence our current estimate.
- Sunshade: 100 kg.
  Scaling the ISO sunshade by the cube of its dimensions gives 76 kg.
- Service Module and equipment platform: 800 kg.
  On ISO this is $\sim$ 600 kg. It is assumed that this will vary only a little with the pure size of the telescope as its functions are similar. Additional mass of attitude control propellant is assumed, as appropriate for a long-duration mission.

TOTAL SATELLITE MASS: 4300 kg.

6.3. OVERVIEW

We end with a picture of the conservatively-designed EDISON mission as a stubby satellite 4.5 m in diameter and 8 m long overall (including service module) with mass somewhat less than 5000 kg, containing a 2.5 m Cassegrain telescope which stabilises between 30 and 40 K around 200 days after launch. Its secondary mirror will be markedly colder. The telescope carries a 100 kg instrument package and occupies a 100,000 km orbit from which its 10+ year lifetime would give many iterative chances to access the whole sky. EDISON is currently under study as an international mission, perhaps to be developed as a joint NASA/ESA "Explorer + M3" venture (see section 8.2).

7. Beyond Earth Orbit: The Promise of L2

The presence of earth moving in the satellite's sky forces many constraints upon the designer. We cannot fully exploit the possibility of cooling from the anti-sun face of the outer surface, pointed out in section 2.3, by simply painting it black, as earth IR and, especially, albedo, would thereby be absorbed in excessive amounts. The telescope aperture must also be protected from earthlight with an aperture shield, and the inner face of such a shield is visible to the telescope tube itself. Although the shield would be self-cooled, and the "visible" surface has a reflective coating with low emissivity, appreciable heat radiation thereby enters the tube.

Most of these drawbacks are removed if we leave the immediate vicinity of the earth. In heliocentric orbit very large distances make transmission of data, even if it is substantially preprocesssed by powerful onboard systems, a difficult problem. The combined considerations make a halo orbit around the Lagrangian point L2 a most attractive option. Attainable with significantly less energy than is needed to enter a circular HEO, such an orbit keeps the satellite within 26 degrees of the sun-earth line, so that it would be straightforward to keep both bodies behind a single shade. This makes it possible to consider all sorts of radical designs for the improved cooling of the telescope.

It should be noted that major gains in performance from the relaxation of design constraints on leaving earth orbit are not in general available to cryogenic missions. No matter what their destined orbits, such telescopes must be cooled and stored before launch and protected when in the vicinity of the earth after launch. Their designs can never, therefore, escape from the confines of the vacuum jacket which must enclose the cryogens at these times. Performance gains for such missions at L2 are accordingly confined to those resulting from the drop in surface temperature caused by the removal of the modest earth-heating effects which remain present in HEO (an impression of the amplitude of these effects can be gained from the of the cyclic temperature variations in Figure 4) and from freedom to optimise the surface properties of the antisun side of the spacecraft.

However, without the periodic intrusion of Earth illumination from the anti-sun side, much more powerful designs for *passive* cooling become practical. An aperture shield (AES) is no longer required (see section 3.3), which removes both the consequent parasitic heating and the need for the launcher to accommodate a structure appreciably larger in diameter than the outermost radiation shield. More important is the opportunity to add additional radiators to utilise the un-illuminated rear side of the satellite for cooling. Figure 10 shows a simple version of the EDISON design which has been modified to exploit these aspects of L2: the radiation shields have been cut away on the antisunward side to define a plane surface for additional radiating area on each shield and on the telescope tube.

Relative to the RCIRT model for high Earth orbit (Section 5 above) this version produces a significant reduction in the expected temperature. We estimate that a realistic model of this configuration will predict optics temperatures in the vicinity of 23K. Such temperatures should enable observations to continue to be limited by astronomical background noise to wavelengths longer than 40 microns, with corresponding enhancement to the telescope performance.

Of such stimulating possibilities is the future of infrared astronomy in the 21st century composed. We intend to investigate them in more detail in the near future.

8. Opportunities in the Next Decade: Small Passively Cooled Missions

8.1. Explorer Class Options and the Japanese SMIRT Mission

The scaling law described in section 3.1 implies that radiatively cooled designs can achieve lower temperatures in small satellites than in large ones. The technique is thus particularly attractive for moderate sized ($\sim$ 1 m) "Explorer" class missions.

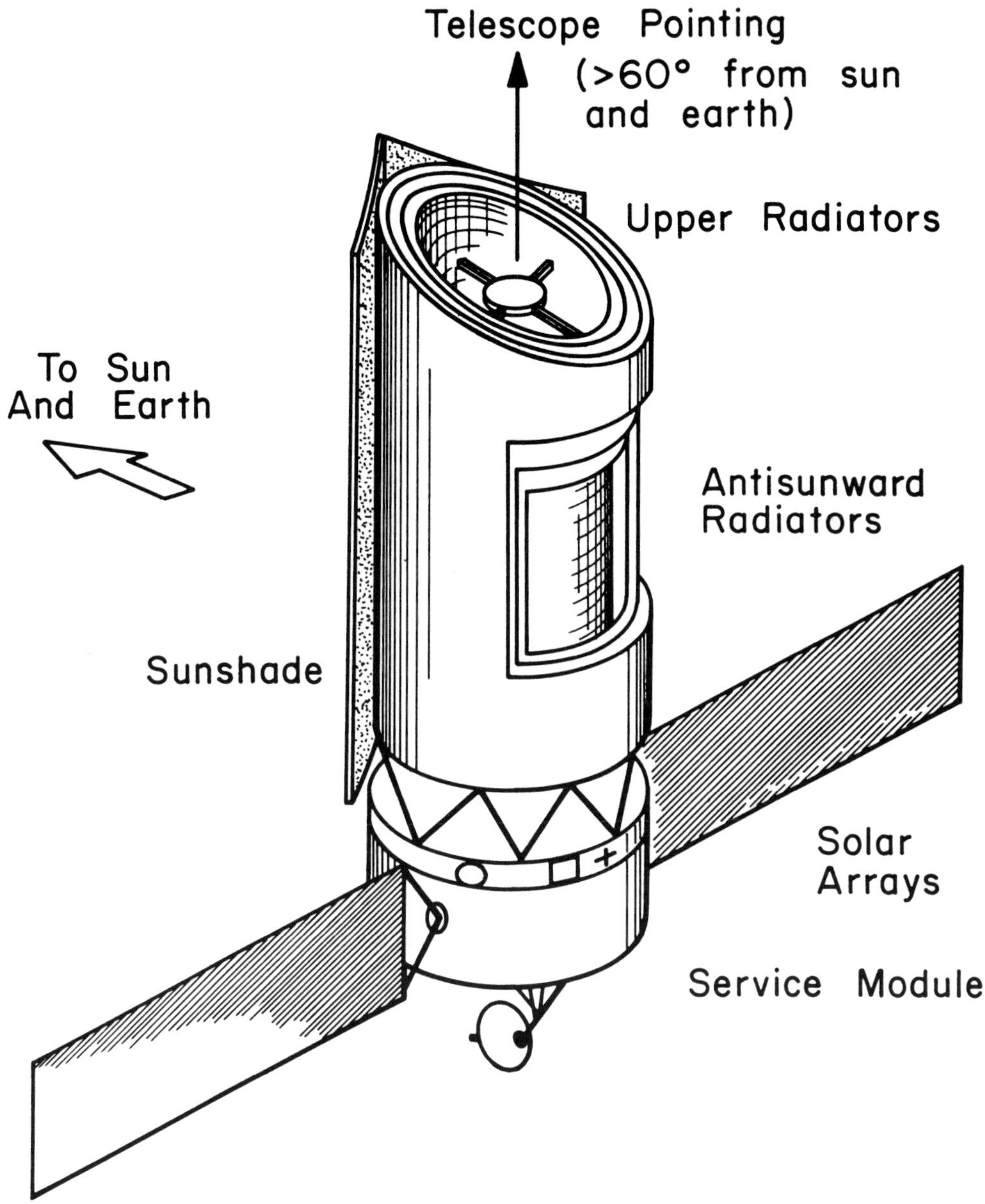

EDISON AT L2

(SCHEMATIC)

Fig. 10. Schematic illustration of a design for the EDISON mission in a halo orbit around the Earth's outer solar Lagrangian point, L2.

The prospect of an infrared successor to the highly successful International Ultraviolet Explorer (IUE) has in fact been a significant motivation for much of the early work of our group. There is a particular attraction in the concept of a long-lived telescope of this type being used to complement shorter-lived cryogenic facilities, which can then be optimised to make maximum use of their lower temperatures for work in the 50 - 150 micron range where their greatest strengths lie (see Fig. 2 of Thronson *et al.* , 1993).

A smaller but potentially very capable mission is in fact being proposed to ISAS in Japan for launch around the turn of the century on the Institute's new M-V launcher. T. Matsumoto and colleagues at Nagoya have developed an outline design for a **SM**all **I**nfra**R**ed **T**elescope (SMIRT), a 1m telescope, diffraction-limited in the near-thermal IR and cooled by a combination of radiation, a closed cycle cooler, and some cryogen boiloff to a temperature in the 20s K, to be placed in a sun-synchronous near-polar orbit from which it would carry out a high-resolution, high-sensitivity sky survey at near- to mid-IR wavelengths.

Although this is not currently planned by the proposers, this mission could readily fulfil the requirement for a high-resolution and -sensitivity survey at 60 microns described by Rowan-Robinson at this meeting as an essential precursor to a new generation of space infrared telescopes. Clearly the survey proposed at Nagoya would be an immensely valuable scientific resource for the early 21st century.

8.2. A Major Mission for the Early 21st Century: M3 + Explorer

The post-submission continuation of the development of the original POIROT proposal (section 5.1) established that constraining the mission to the M2 guidelines reduced its capabilities disproportionately to the savings.

A 25% increase in the mission budget would have allowed a dedicated launch on an Ariane 4. Davies *et al.* (1991) have outlined the major gains in the capability of the satellite which could thus be achieved. A single launch would permit the placement of a satellite of up to $\sim$ 1700 kg in a 72 h elliptical orbit with perigee height of 20,000 km (though the increased mass could probably not be fully exploited without a further increase in resources). At this altitude earth heating at perigee is substantially reduced (the earth's solid angle is reduced by a factor of 16, and perigee passes are 3 times less frequent than in the ISO 24 h orbit) while much of each orbit would be spent above 100,000 km (the apogee altitude is $\sim$ 140,000 km), so an average surface temperature lower than that of SIRTF could be expected. The higher perigee would allow the size of the AES to be greatly reduced, permitting it to be accommodated in a standard Ariane fairing. This orbit also places perigee well above the main radiation belts, a *bête noire* of infrared detectors. Furthermore, for an Ariane 4, a single-payload launch allows a reduction in the minimum axial frequency from 31 to 18 Hz (Loftus, Texeira and France, 1991) and consequently a reduction by a factor of up to 3 in the sizes and conductivities of the support straps. A significantly lower final temperature would therefore result, both from the reduction in average earth heating and from this reduction in strap conductivity. Seldom can so small a budget increase have offered such cost-effective increases in a spacecraft's potential performance!

However even greater gains can be foreseen if a POIROT-type mission were scaled to the resources which would result if the ESA Medium Mission could be more evenly matched by a partner. The result would be access to a launcher large enough to deliver a payload of approximately 3000 kg to a halo orbit about L2 by a lunar slingshot manouevre, as described by J. B. Farrow at this workshop. As pointed out in section 7, L2 offers great attractions for the deployment of aggressively designed radiatively cooled telescopes. This level of resources would also suffice to exploit the larger deliverable mass.

We are therefore investigating the possibility of a joint ESA/NASA venture in designing, developing and launching a modestly smaller EDISON spacecraft than outlined in section 6. At this time we envisage a combination of an ESA Medium Mission with a NASA Medium Explorer, which could allow an extremely capable 2 m+ radiatively-cooled telescope to be built and launched.

Acknowledgements

We thank a number of friends and colleagues for enthusiastic participation which contributed much input and stimulus, and also for their patience as ideas evolved into suggestions and even into proposals. Mark Casali very kindly wrote the Monte-Carlo simulator which proved invaluable for investigating thermal radiation fields near curved surfaces, a process which had defeated our analytical abilities. Masahiko Hayashi acquainted us with the plans developing around the M-V launcher in Japan, and Professor T. Matsumoto kindly communicated, and allowed us to discuss, details of the hybrid mission being designed in Nagoya. At the Edinburgh workshop, too, John Farrow drew our attention to the possibilities of L2 and other orbits which we had previously thought of as highly exotic. Subsequently, John Bally made certain that we thought hard about their implications, and several of his ideas are indorporated in this paper. John Mather and Max Nein allowed us access to the results of preliminary studies of related missions. We are particularly indebted to the Technology and Computing Division of the Royal Observatory, Edinburgh, and in particular to Donald Pettie, for extensive assistance and support, especially with early models of POIROT. Acknowledgement is also made to NASA grant NAG 8–899. TGH thanks Frances Hawarden for her consistent encouragement and support, and HAT acknowledges the material assistance of Tina Thronson in financing this work.

References

Bramson, M.A.: 1968, *Infrared Radiation, a Handbook for Applications*, trans. Rodman, R.B., ed. Wolfe, W.L., Plenum, NY

Davidson, A. M., Seidel, A. and Wanner, M.: 1989, *ESA Bulletin* **57**, 53

Davies, J. K., Hawarden, T. G. and Mountain, C. M.: 1991, *Acta Astronautica* **25**, 223
Emery, R. J.and Burgio, C.: 1982, *Infrared Space Observatory: Report on the Phase A Study*, ESA SCI(82)6
Gautier, T.N., Boulanger, F., Pérault, M. and Puget, J-L.: 1992, *Astron. J.* **103**, 1313
Helou, G. and Beichman, C. A.: 1990, *Proc. 29th Liege Intl. Astrophys. Colloq.*
Hust, J. G.: 1983, in R. P. Reed and A. F. Clark, ed(s)., *Materials at low temperatures*, American Society for Metals/National Bureau of Standards, 133
LaRocca, A.J.: 1978, in W.L. Wolfe and G.J. Zissis, ed(s)., *Infrared Handbook (2nd ed.)*, US Govt. Printing Office,
Lee, J. H.: 1990, *Cryogenics* **30**, 166
Loftus, J. P., Texeira, C. and France, M. E. B.: 1991, in J. R. Wertz and J. L. Wiley, ed(s)., *Space Mission Analysis and Design*, Kluwer, 607
Mason, P. V.: 1988, *Cryogenics* **28**, 137
McMordie, R. K.: 1991, in J. R. Wertz and J. L. Wiley, ed(s)., *Space Mission Analysis and Design*, Kluwer, 370
Orlowska, A. H., Bradshaw, T. W. and Hieatt, J.: 1990, *Cryogenics* **30**, 246
Sparks, L. L.: 1983, in R. P. Reed and A. F. Clark, ed(s)., *Materials at low temperatures*, American Society for Metals/National Bureau of Standards, 47
Thronson, H. A., Jr., Davies, J. K., Hackwell, J., Hawarden, T. G., Knacke, R. F., Lester, D. F. and Mountain, C. M.: 1993, *Space Science Reviews* , this volume
Williamson, M.: 1987, *Phys. Technol.* **18**, 120

EDISON: THE NEXT GENERATION INFRARED SPACE OBSERVATORY

H. A. THRONSON, JR
Wyoming Infrared Observatory, Campus Box 3905, University of Wyoming, Laramie, WY 82071, U.S.A.
and
Royal Observatory, Blackford Hill, Edinburgh EH9 3HJ, U.K.

J. K. DAVIES
Royal Observatory, Blackford Hill, Edinburgh EH9 3HJ, U.K.

J. HACKWELL
The Aerospace Corporation, P. O. Box 92957, MS M2-266, Los Angeles, CA 90009, U. S. A.

T. G. HAWARDEN
Royal Observatory, Blackford Hill, Edinburgh EH9 3HJ, U.K.

R. F. KNACKE
State University of New York, Dept. Earth and Space Sciences, Stony Brook, NY 11794, U. S. A.

D. LESTER
University of Texas, Dept of Astronomy, RLM Hall, Austin, TX 78712, U. S. A.

and

C. M. MOUNTAIN
Royal Observatory, Blackford Hill, Edinburgh EH9 3HJ, U.K.

(Received 19 September, 1991)

Abstract. EDISON, a large-aperture, radiatively-cooled telescope, is proposed as the major international mission to follow the current generation of cryogenically-cooled infrared space telescopes. It is being studied at present as a 2.5–3.5 m mixed radiatively- and mechanically-cooled facility optimized to investigate the wavelength range 3–100+ μm. This paper outlines the status of the project, discusses some aspects of a smaller-aperture 'precursor' mission, and describes a portion of the baseline science mission.

1. The Infrared Universe and First Generation Space Observatories

So long as observations in the infrared (1–500 μm) are limited, so too will be our understanding of the cosmos. This is a direct consequence of the richness of this wavelength regime: critical diagnostic features of the solid state and gaseous interstellar medium (ISM) are found in the infrared, as are photospheric or atmospheric features of planets, brown dwarfs, and other sub-stellar objects. Supernovae and novae, enriching the ISM with heavy elements, can be studied via their infrared spectra. Finally, the expansion of the universe shifts major visual diagnostic lines

Space Science Reviews **61:** 145–169, 1992.

into the infrared, along with the majority of the continuum emission from very distant objects.

Ground-based observations provide important and tantalizing glimpses of the infrared universe. However, even outstanding sites such as Mauna Kea and Antarctica are seriously limited by atmospheric absorption at wavelengths longward of about 2.5 μm. Furthermore, even if the atmosphere was transparent, the emission from faint celestial sources is overwhelmed by that from the bright warm telescope.

Motivated by the richness of the infrared and recognizing the limitations of ground-based observatories, astronomers have used telescopes carried aloft in balloons, airplanes, and rockets. Most notable among these observatories has been the 0.9 m telescope onboard NASA's Kuiper Airborne Observatory (KAO), which is planned to be surpassed this decade by the 2.5 m telescope of the Stratospheric Observatory For Infrared Astronomy (SOFIA). Later this decade, one of the instruments on the Hubble Space Telescope (HST) will be replaced with the Near-Infrared Camera and Multi-Object Spectrometer (NICMOS).

Ultimately, however, the most sensitive infrared observatories have been launched into space and cooled via liquid helium. The US/Dutch/UK Infrared Astronomical Satellite (IRAS) conclusively demonstrated the technical mastery of cryogenic telescopes in space and further underscored the importance of observations at these wavelengths. More recently, the Cosmic Background Explorer (COBE) produced fundamental information about the relic radiation from the Big Bang, its distribution on the sky, as well as the large-scale infrared emission from the Milky Way. The Infrared Space Observatory (ISO) will be far more sophisticated and sensitive than IRAS and is planned for launch by the European Space Agency in 1993. At about the same time, the Japanese will launch the small aperture Infrared Telescope in Space (IRTS) to undertake a survey of diffuse emission from the interstellar medium. Finally, about a decade from now, NASA plans to launch the Space Infrared Telescope Facility (SIRTF), by far the most powerful of these cryogenic infrared space telescopes and the finale to NASA's Great Observatories programme.

Current and planned generations of infrared observatories owe their tremendous sensitivity improvements over groundbased telescopes to cryogenic cooling, making the telescope emission extremely low, and to the advancement in detector technology over the past decade. However, theoretical limits to some detector performance characteristics are already in sight. As these limits are approached, further major improvements in sensitivity will depend more upon increasing telescope aperture. The sensitivity of astronomical observations – here defined as the inverse of the integration time required to reach a given flux level – increases as a high power of the aperture: sensitivity $\propto$ (diameter)$^{2\rightarrow 4}$ depending upon such things as background emission, contribution from the variety of noise sources, and source size. Cryogenically-cooled observatories are limited in aperture by their large tanks of liquid helium. ISO and SIRTF, for example, already approach the maximum size and mass that can be launched by rockets available to Western space agencies in the coming decade. Therefore, to exploit fully the advantages of operation in space via large apertures, an alternative to cooling solely via cryogen tanks will be necessary for future missions.

In the US, the National Academy of Sciences Astronomy and Astrophysics Survey Committee recognized that new techniques and technology will lead to more powerful space observatories for the entire electromagnetic spectrum. In its recent report, the committee called for increased support for development of future technology. Success in this development, the report expects, will lead to the choice of post-Great Observatory missions before the end of this decade. At the same time, the committee also recommends international missions for those most ambitious programmes that may exceed the resources available within a single country or agency.

EDISON is proposed as an international infrared space observatory to follow and build upon the first generation of cryogenic missions. Using radiative cooling, this facility will not require the large tanks that limit the aperture of telescopes cooled by expendable cryogens. Based on our present understanding of the structure of a radiatively-cooled telescope, we estimate that Western launch vehicles available later this decade will be able to carry a telescope of this design with an aperture as large as about 3 m. Such an increase in light-gathering power will be a major advance in sensitivity and angular resolution. At the same time, it may be worthwhile to develop modest-aperture designs for radiatively-cooled telescopes, Explorer-class facilities which would be precursors to EDISON as described in this paper.

2. Sensitivities of Passively-Cooled Telescopes in Space

Groundbased telescopes have taken advantage of radiative cooling for decades, with modest increases in sensitivity in the mid-infrared windows during winter observing runs as the telescope temperature and, consequently, its infrared emission drops. The most effective terrestrial site for this technique will probably be the Antarctic, where an infrared observatory is proposed for construction during the 1990's. However, major cooling of groundbased telescopes is severely limited by thermal conduction to the Earth and ultimately by freezing of the atmosphere onto the structure. Truly effective cooling of telescopes, whether via cryogens, mechanical refrigerators, radiation, or some combination, can only take place in space.

Cryogenically-cooled satellites, such as ISO and SIRTF, will be the most sensitive telescopes of their size throughout the infrared. In their designs, they have traded aperture and longevity to achieve the extremely low temperatures necessary to reduce the telescope emission for celestial background-limited observations in the far-infrared ($\lambda \geq 40$ μm). In contrast, to maintain large aperture and long lifetime, even very effectively radiatively-cooled facilities, such as EDISON or a lunar polar telescope (Lester, 1991), accept about an order of magnitude higher telescope temperature and a significant increase in non-celestial emission at the longest infrared wavelengths. This higher telescope emission in the far-infrared is, however, partially compensated for by the larger aperture possible for radiatively-cooled satellites.[1] First generation (purely cryogenic) satellites have the additional advantage that the liquid helium which cools the telescope is also used to cool the detector systems. For equally good sensitivity at long wavelengths, our current design for EDISON may require a mixture of higher-temperature detectors, mechanical refrigerators, and/or modest amounts of cryogens. These have been identified as major technical hurdles for the radiatively-cooled missions that we are studying.

The outstanding near- (1–5 μm) and mid-infrared (5–40 μm) performance of radiatively-cooled space telescopes, whether in orbit or on the moon, can be demonstrated by considering one likely type of observation, high-spatial-resolution broad-band observations (including imaging) of point sources. In the following discussion, it is assumed that the optical designs of all telescopes are the same, with sufficiently low-noise detectors that all observations are limited only by the background (telescope or celestial). Telescope emission is assumed to be $\epsilon = 0.1$ in all cases, with $\lambda/\Delta\lambda = 10$, diffraction-limited pixels, and an integration time of 1 second for the hypothetical instruments. We emphasize that current technology detectors, especially at near-infrared wavelengths, are still far from reaching the celestial background toward the darkest parts of the sky.

Figure 1 presents estimates of the noise from celestial emission and the telescope for three missions: a 1 m cryogenic facility with T_{tel} = 3 K, a 6 m "post-HST" satellite optimized for UV and visual observations with T_{tel} = 100 K, plus EDISON, which we take to be a 3 m telescope with T_{tel} = 40 K. Emission from a $\sim$ 300 K ground-based telescope would be so strong as to be far off-scale in this figure in the range $\sim$ 5–200 μm. In any case, observations from the ground are restricted to only a few narrow windows between 3 and 300 μm.

Performance of these facilities can be discussed by deriving relative signal-to-noise ratios, which under the assumptions here is proportional to the square of the telescope diameter divided by the largest noise source (Fig. 1) for each telescope. This is presented in Figure 2, where we include two options that we are evaluating for the orbit of EDISON: a generic "1 AU" location such as L2, as well as the interesting alternative of a trans-asteroid orbit, roughly 4 AU, beyond the emission and scattered light of the zodiacal dust.

Going from the shortest to the longest wavelengths in Figure 2, we can evaluate strengths and weaknesses of the missions under the assumptions adopted here. Bear in mind that we have plotted the logarithm of the signal-to-noise, which for background-limited observations depends upon the length of integration as $(\text{time})^{\frac{1}{2}}$. Consequently, signal-to-noise values that differ for a pair of telescopes by,

[1] EDISON, at a temperature of $\sim$ 40 K, will of course be celestial background-limited at near- and mid-infrared wavelengths, if future detector technology allows.

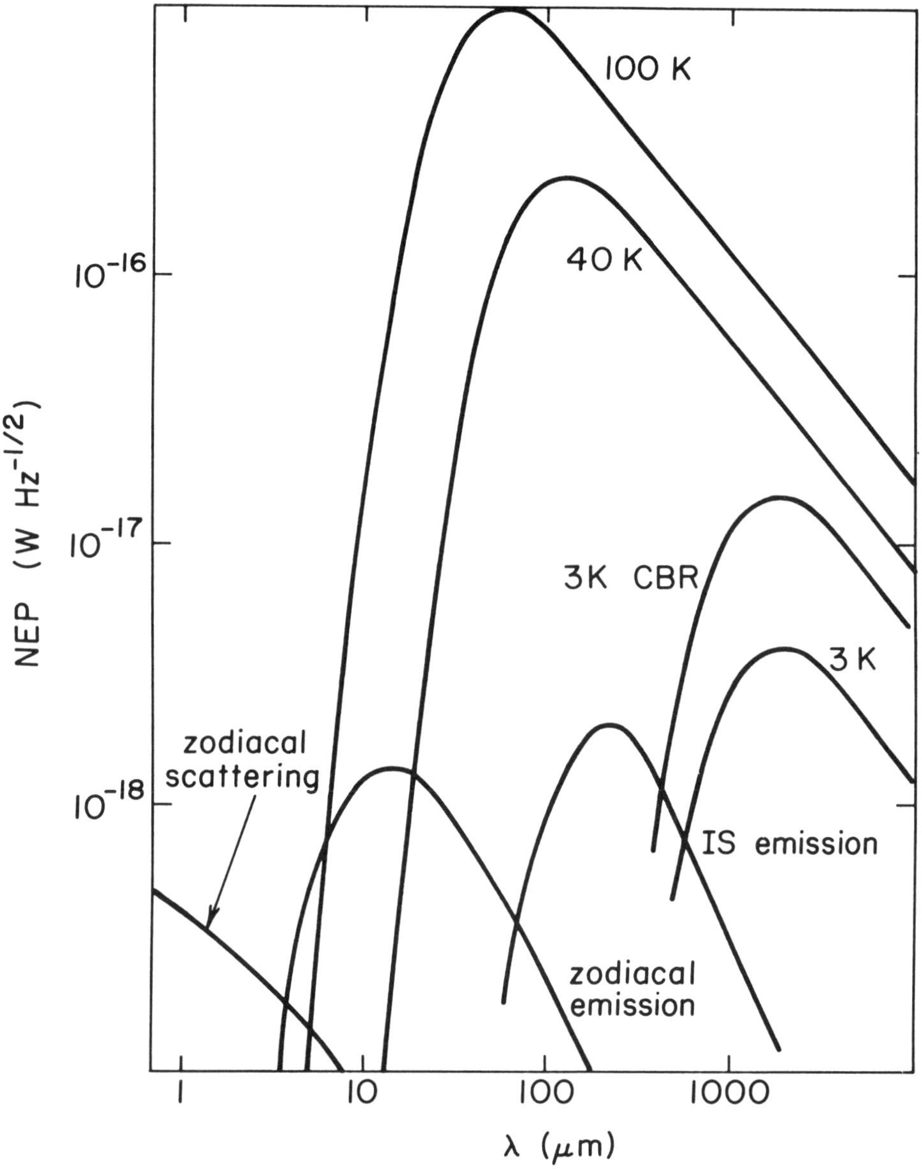

Fig. 1. Celestial and telescope noise contributions to infrared photometry and imaging for three future space missions with identical optical designs, spectral resolutions of $\lambda/\Delta\lambda = 10$, and integrations times of 1 second. The zodiacal emission and scattering are for the ecliptic pole from a location near the Earth and the interstellar dust emission is appropriate to high galactic latitudes. Observations of this type with current technology will be dominated by detector noise at shorter wavelengths. Confusion effects can limit the sensitivity of smaller aperture telescopes at longer wavelengths. Neither effect is included in this figure.

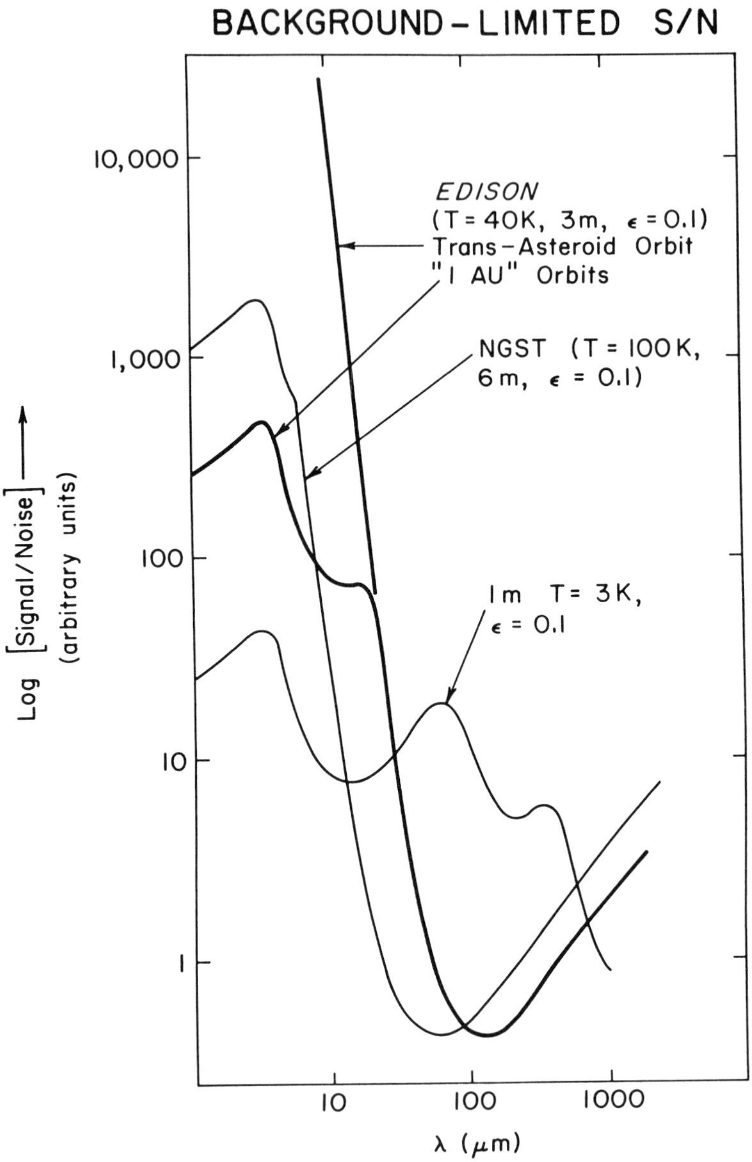

Fig. 2. Logarithm of the relative signal-to-noise ratio for photometry (and imaging) under the assumptions described in Figure 1 for three space missions operating at the telescope and celestial background limit. The signal-to-noise ratio here is the square of the telescope diameter divided by the largest noise source for each facility from Figure 1. The vertical units are arbitrary. The performance of EDISON is shown both for a location in the vicinity of the Earth ("1 AU" orbit) and a location beyond the zodiacal dust. The same telescope equilibrium temperature (40 K) is adopted here for both locations. Except for the shortest and longest wavelengths, even the largest groundbased telescope is off the bottom of this plot. Detector noise and the effects of confusion are assumed to be negligible.

say, a factor of 10 in Figure 2 require integration times a factor of 100 longer. The figure greatly understates the advantage of infrared observations in space by any satellite by not including ground-based telescopes for comparison. As an example, consider SIRTF, which with its very low temperature was especially optimized for operation at far-infrared wavelengths. Despite this, even at shorter mid-infrared wavelengths, around 10–20 μm, SIRTF should reach a signal-to-noise about $1\frac{1}{2}$ orders of magnitude greater than a ground-based 8 m ambient temperature telescope for the same integration time. This translates into a decrease in integration time by a factor of 1,000! This tremendous advantage of even modest-aperture space observatories, coupled with the inaccessibility of most of the infrared from the ground, is the overwhelming justification for infrared missions in space.

At the shortest wavelengths considered here, Figure 2 shows that the missions are all limited by scattered zodiacal light as even the warmest of the three baseline telescopes emits little or nothing in the near-infrared. In this regime, aperture counts for everything, with the 6 m telescope of course the most powerful for the point source photometry that we are describing here. For orbits inside the zodiacal dust belt, the maximum signal-to-noise ratio is reached near 3 μm, the darkest part of the sky from the inner solar system, where the falling scattered zodiacal light emission crosses the rising thermal flux from the same material (Fig. 1). Slightly longward of this wavelength, the emission from the 100 K telescope begins to become significant, the signal-to-noise ratio of this facility decreases precipitously, and EDISON becomes the most sensitive infrared telescope throughout the mid-infrared.

In Section 4 we outline some of our considerations concerning moderate-aperture ($\sim$ 1–2 m) radiatively-cooled Explorer-class precursors to EDISON. Performance capabilities of such missions can be estimated from the two figures by simple scaling. Figure 1 is entirely independent of aperture, while the results in Figure 2 scale as the square of the primary diameter. For example, for the same optical design and detectors, both 1 m radiatively-cooled (T $\approx$ 40 K) and 1 m cryogenically-cooled (T $\approx$ 3 K) telescopes perform equally well throughout the near- and most of the mid-infrared as these wavelengths are well down on the Wien side of the telescope emission.

Figures 1 and 2 show why we are investigating the trade-offs in trans-asteroid orbits (TAO) for EDISON. Beyond the zodiacal dust, which is believed to be largely confined to the asteroid belt, observations are limited by composite light from distant stars and galaxies, plus emission from the telescope, which for a 40 K system is extremely weak. For example, our simple analysis indicates that for the same integration time EDISON would reach a signal-to-noise ratio at 10 μm roughly *300,000 times greater* than an 8 m 300 K telescope on the ground if both work at their background limits. For noise decreasing as the square root of the time, an 8 m ground-based telescope would have to integrate *90 billion times longer* than EDISON to reach the same signal-to-noise![2] Obviously, in any sense these numbers are so huge as to be meaningless and our simple assumptions break down. For example, the necessary extremely low-noise detectors are beyond present technology. It is sufficient here to emphasize that no presently conceivable ground-based telescope can compare with the broadband sensitivities achievable in the near- and mid-infrared by EDISON, to say nothing of the great losses in this regime due to absorption by the atmosphere. However, as Figure 1 shows, the equilibrium temperature of any infrared telescope in TAO must get well below $\sim$ 100 K, as EDISON will, to take advantage of a major reduction in the solar system dust emission.

Note that for this simple comparison, we have adopted the same temperature for EDISON in both inner and outer solar system orbits, although we expect the equilibrium temperature to be significantly lower than 40 K if the facility is at a distance of 4 AU from the Sun. Lower equilibrium temperatures will not improve the performance of the facility at wavelengths *much* shorter or longer than the Planck maximum, of course (Fig. 1), but the potential far-infrared ($\sim$ 50–200 μm) sensitivity of EDISON will increase.

Near the maximum of the blackbody emission from telescopes, gains in sensitivity are possible by decreasing the temperature, as an alternative to a large light-gathering aperture, as Figures 1 and 2 demonstrate. At far-infrared wavelengths, the emission from the two radiatively-cooled facilities becomes significant and their estimated sensitivities would be modest were it not for partial compensation by large apertures. It is in the far-infrared that cryogenic telescopes may have clear advantages for sensitive photometry and broadband imaging. However, at long wavelengths the limits to sensitivity may not be celestial or telescope thermal emission after all.

Approaching sub-millimeter wavelengths ($\lambda \sim 100$ μm), broadband sensitivity calculations become uncertain due to the effects of confusion, which we did not include in our figures. Helou and Beichman (1991) and Gautier *et. al.* (1992) estimated that spatial structure both in the infrared "cirrus" and in the light from distant galaxies will produce an irreducible limit to sensitivities which are strong functions of aperture. It was for this reason, among others, that we have chosen to maximize aperture which has the effect of increasing system sensitivity in a confusion-limited regime. Our preliminary estimates demonstrate that even at the maximum of EDISON's thermal emission, achievable sensitivities are much higher in a confusion-limited environment than is the case for smaller cryogenic satellites.

Figure 2 demonstrates why the large-aperture, radiatively-cooled sub-millimeter missions such as FIRST (Frisk, 1991) or SMM (Phillips, 1991) have adopted much the same philosophy as EDISON has in the near- and mid-infrared: emphasis on

[2] For comparison, this difference in time is comparable to the difference between one beat of a bat's wings and a human lifetime. Even if this estimate is optimistic by a factor of one million, in a few minutes of integration EDISON will reach a noise level similar to that achievable by an 8 m ground based telescope after integrating for its expected lifetime.

aperture at the expense of higher telescope temperatures than cryogenic missions.

3. The EDISON Infrared Space Observatory

EDISON[3] is in an early stage of study and assessment. As a consequence, the present design must be considered preliminary, little optimization has yet been undertaken, and only limited systems analysis has been possible. Nevertheless, basic design goals – that of large aperture and long lifetime, as examples – are clear. Preliminary work on optical design, thermal modeling, closed-cycle cooler requirements, orbits and launch vehicle capabilities, and satellite subsystems is discussed elsewhere in this volume.

At present, EDISON is being evaluated as a simple large-aperture Cassegrain telescope, surrounded by a set of concentric radiation shields, although other optical designs have been considered (Humphries *et al.*, this volume). Part of our evaluation procedure includes alternative designs, including "off-axis" telescopes and more sophisticated radiation shields.

Some telescope and spacecraft parameters are summarized in Table I.

3.1. Telescope Aperture and Spacecraft Size and Mass

Given the strong dependence of sensitivity on aperture, an overarching goal of our design is to maximize telescope light-collecting area to the limit allowed by available launch vehicles. From the results of the thermal models to date, the full diameter of a radiatively-cooled satellite is about 1.5 times the telescope aperture (see section by Hawarden *et al.* in this volume), excluding critical appendages such as the solar array/sun shield and antennae. The available space within a standard upper stage fairing of a Titan IV – the western world's most powerful expendable launch vehicle – allows a maximum telescope diameter somewhat larger than about 3 m using this criterion. This is more than an order of magnitude greater light-gathering capability than the largest cryogenic space observatory so far planned.

With the telescope and surrounding support and radiating structures nearly filling the launch vehicle upper stage, even a simple solar cell array/sun shield may have to be deployable. The concept of a lightweight, deployable sunshade is being investigated by the technology development program for the ESA FIRST mission, for example. One-third scale models of an inflatable thermal shield that can be erected and rigidized in orbit have already been demonstrated (Frisk, 1991).

The size and mass budget for EDISON have by no means been optimized, but only estimated to allow preliminary study. We have constrained the total payload

[3] An historical note: Thomas Alva Edison can be fairly described as the outstanding infrared astronomer of the last century. In 1878, using a detector of his own design and construction, Edison detected the infrared emission from both α Boo and the solar corona during an eclipse visible in Wyoming. He also devised the bolometer, which is described in correspondence with Samuel Langley, although Langley discouraged Edison from pursuing the idea.

TABLE I
EDISON Design Goals and Spacecraft Characteristics
[Outline; Autumn 1991]

TELESCOPE	
Aperture	2.5–3.5 m
Primary construction	monolithic
Primary mass	$\sim$ 1000–1500 kg
Primary f/ratio	f/1.2
Telescope f/ratio	f/8–f/16
Field of view	$2'$–$5'$ per instrument
Tertiary	fixed pyramid ($\geq$ 4 faces?)
Angular resolution	diffraction limited at all wavelengths longward of $\sim 5\ \mu$m
Tracking accuracy (3 m aperture)	$0.''03$ - $0.''06$ (1σ rms for several hours)
Operating temperature (primary)	$\leq$ 50 K
Total telescope absorption efficiency	$\geq$ 0.9 after 5 yrs
DETECTORS and INSTRUMENTS	
Spectral coverage	2–100+ μm
Spectral resolution	$\lambda/\Delta\lambda \sim 5 \rightarrow 10^{4+}$
Imaging requirements	$0.''2$ pixel^{-1} at 6 μm
Detector cooling	$\sim$ 2–5 μm: Radiative (T $\sim$ 40 - 70 K) $\sim$ 5–15+ μm: Refrigerator or hybrid (T $\sim$ 5–15 K) $\sim$ 15–40+ μm: Refrigerator or hybrid (T $\sim$ 2–4 K)
SPACECRAFT	
Orbit	L2 halo, trans-asteroid orbit, or high circular orbit
Size (incl. service module; not incl. sunshield)	5 m $\times$ 8 m
Total mass (approx; limited by Titan IV launch)	$\leq$ 5000 kg
Suspended mass	2500–3000 kg
Cold mass	1500–2500 kg
Lifetime	10+ yrs
Total power requirements	$\sim$ 1–5 kW
Data acquisition rate	$\sim$ 1 mbps

mass to be less than 5000 kg and for the entire spacecraft to fit within a cylinder of diameter 5 m and length 8 m. This size estimate is within the envelope of a large Titan IV fairing and the mass is less than the capacity of the block change Titan IV (planned for the mid-1990s) to lift to high circular orbit (HCO), L2 halo, or a trans-asteroid orbit (TAO) with gravity assists. For comparison, SIRTF will have a mass of about 4500 kg, with an overall external size somewhat smaller than that of EDISON.

The current partial mass budget is included in Table I, with other discussion of the values chosen for the cold mass by Hawarden *et al.* in this volume. Early thermal modeling for EDISON adopted some results for the designs for ISO, SIRTF, SOFIA, and the Orbiting Planetary Telescope (OPT). Although most of the mass estimates fall in the range estimated for somewhat similar structures in other spacecraft, we expect significant development in lightweight structures will be necessary.

3.2. Optical Design and Tracking Requirements

The current design goal for EDISON is diffraction-limited images at 3–6 μm. This is about the shortest wavelength for which the space telescope will be much more sensitive than is possible for any ground-based telescope. This band is also near the location of the darkest wavelength of the celestial sky from orbits in the inner solar system. A much more detailed discussion of some optical designs for EDISON is given by Humphries *et al.* elsewhere in this volume, but our optical design goal is an angular resolution of $1.22\lambda/D = 0.''5$ at 6 μm for a 3 m primary. Because the optical performance of EDISON translates into demanding pointing and tracking requirements, we shall evaluate carefully the scientific justification for high performance at the shortest wavelengths at which the facility is expected to operate.

Maximum spatial information will be possible with about 5 resolution elements of some imaging device over the diffraction pattern. At 6 μm, the angular distance across the first minimum in the diffraction pattern is $1''$, leading to a requirement of $0.''2$ pixel^{-1} for EDISON's hypothetical shortest-wavelength imaging device. This performance translates directly into requirements for tracking, which at this early stage we approximate as 1/3 of a pixel in the imaging device or $0.''06$ (1σ rms) if we adopt 6 μm as the shortest wavelength of diffraction-limited performance. Our analysis has not yet proceeded to the point where we can estimate the length of time over which this accuracy must be held, although we assume that it must be hours to allow long exposures.

For comparison, the current fine lock tracking ability of HST is approximately $0.''008$ (1σ rms) for a few hours, which we adopt as representative of the current level of technology. The recent ESA/NASA/DARA study of the Orbiting Planetary Telescope concluded that its tracking requirement of $0.''02$ was feasible given current understanding of this decade's technology. However, all system elements in OPT would be required to meet stringent requirements and focal plane tracking within some instruments may be necessary. Therefore, we conclude that our present baseline for EDISON is demanding, but well within the capabilities of near-future technology.

Tracking by EDISON will be a major issue, as it is for all space astronomy facilities. However, since the severe requirement outlined above is useful only at the shortest wavelengths proposed for the satellite, as we further develop the science requirements, it may be possible to relax the demands on the tracking with no sacrifice in scientific capability.

3.3. Detectors and their Cooling Systems

3.3.1. Future Detector Goals

Although passive cooling in space is required for long-lived, large-aperture infrared telescopes, temperatures achievable in the focal plane by radiative cooling alone may to be too high for effective operation of current generation infrared detector systems at wavelengths longer than the near-infrared. Recent promising developments in closed-cycle cooler technology are described elsewhere in these proceedings. At the same time, development of sensitive detectors that operate at higher temperatures than those planned for cryogenic missions may allow wider wavelength of operation via radiative cooling alone. "High temperature" infrared detectors should also ease the requirements on potential mechanical coolers and, perhaps, extend mission lifetime if cryogens must be used for some telescope systems.

Higher temperature infrared detectors have been identified by, for example, NASA's 1991 Astrotech 21 Technology Workshop as a major goal for the coming decade. For the purpose of outlining the EDISON mission at this early stage, we assume that satisfactory near-infrared ($\sim$ 1–5 μm) detectors will be available, as significant progress has been made for HST's NICMOS instrument. Its HgCdTe photovoltaic detectors are designed for work at about 60 K, apparently easily achievable via radiative cooling alone. In the mid-infrared ($\sim$ 5–30 μm), doped-silicon impurity-band-conduction (IBC) sensors are planned for SIRTF. This technology has the potential for higher temperature operation, perhaps to the temperatures reached by either radiative cooling in orbit (T $\sim$ 40–50 K) or reliable two-stage coolers (T $\sim$ 20 K). In addition, there has been much work on "warm" HgCdTe detectors for 1–10 μm which might operate even more sensitively at the temperatures achievable by EDISON.

Here we briefly contrast EDISON with the hypothetical future missions considered by the Astrotech 21 Sensor Technology Panel. They assumed, naturally, that post-ISO/SIRTF infrared missions will be radiatively cooled, but since so little thermal design has taken place and been described in the literature, equilibrium temperatures assumed were in the range of 80–150 K. Were this the case, detectors operating broadband in the mid- and far-infrared will probably be in a high background environment. In contrast, the design goal for the EDISON primary mirror, supported by our thermal modeling, is 40–50 K. Consequently, the near- and mid-infrared detectors will operate in a low-background environment, even in broadband mode (Fig. 1), and the instruments should equilibrate at much lower temperatures than anticipated based on past literature on radiative cooling. Furthermore, EDISON's spectrometers will be low background throughout their operating wavelength range. Similarly, an infrared telescope located in a sunless crater at the lunar poles should equilibrate at temperatures in the range of 20–30 K (Lester, 1991), so this far-future mission also requires detectors operating in a very low background environment.

Although sensitive near- and mid-infrared detectors for operation at radiatively-cooled temperatures ($T \geq 40$ K) seem a promising direction for technology development over the coming decade or so, far-infrared systems ($\lambda \geq 40\ \mu$m) are likely to require additional cooling. Typical goals for future sensitive long-wavelength detectors are operating temperatures in the range of 2–4 K. Doped-germanium blocked impurity band (BIB) detectors have been specifically identified by the 1991 Astrotech 21 workshop as the likely technology for early-21st century far-infrared sensors. As with the shorter-wavelength detectors described above, there may be some technological demands by EDISON that are different from those in the hypothetical space missions considered by the Workshop participants. The most obvious example, again, is that the telescope background for EDISON will be much lower than for generic radiatively-cooled telescopes that are outlined in the literature. The maximum emission from EDISON occurs close to 100 μm, but the flux is orders of magnitude less than that from, for example, room temperature telescopes (see Section 2). At very far-infrared wavelengths, on the Rayleigh-Jeans side of the telescope emission, EDISON's emission is only a factor of several above that of cryogenic telescopes with $T \approx 3$ K (Fig. 1). In consequence, it may be premature to adopt a design goal of "high background performance" for post-ISO/SIRTF infrared detectors.

3.3.2. Cooling Systems

While radiative cooling to 40–50 K may be entirely sufficient for near-infrared and, perhaps, mid-infrared detectors, additional cooling almost certainly will be required for EDISON's instrument systems to operate over the design goal range of 2–100+ μm. At present, we are evaluating two methods of additional cooling to reach temperatures of a few degrees Kelvin necessary for the longest wavelengths: closed-cycle refrigerators or modest amounts of onboard cryogens for "point cooling" of the sensors and small parts of the instruments. Hybrid systems, for example with mechanical refrigerators used to cool jackets which surround the cryogens, are generalizations of the two basic techniques.

The status of non-cryogen cooling systems is described widely in the literature and summarized elsewhere in these proceedings. Long-lived Stirling cycle refrigerators for operation at liquid nitrogen temperatures (60–80 K) have been space qualified for some years and were launched in July 1991 in the Along Track Scanning Radiometer (ATSR) on ERS-1 and on the Improved Stratospheric and Mesospheric Sounder (ISAMS) on UARS in September 1991. These are single-stage systems and reliable two-stage systems using the same principles for cooling below 30 K are due to finish qualification testing soon (Orlowska, Bradshaw, and Hieatt, 1990; this volume). Our current design for EDISON assumes capable refrigerators available for cooling to ~ 20 K if necessary.

These systems form the first stages of several possible refrigerator designs for temperatures of a few degrees Kelvin. The most commonly discussed system has been a Joule-Thomson system, with the space-qualified Stirling cycle refrigerators providing two-stage precooling. Optimistic predictions are for space qualification within a few years.

3.4. Orbits

Spacecraft location is a major factor even in early design considerations for EDISON for several reasons. First, given that the Earth and Sun can be major heat sources, locating an infrared observatory as far as possible from both should lead to a lower telescope temperature and improved long-wavelength performance. Greater distance can also mean increased access to large areas of the sky or more detailed observations for some objects. At the same time, increasing distance from the Earth and Sun may mean greater difficulty in meeting requirements for electrical power, pointing and tracking, telemetry, and consumables.

Three distinctly different locations are under consideration for EDISON which we outline here in order of increasing distance from earth.

1. *High Circular Orbit (HCO).* At a distance of about 100,000 km, just outside Earth's radiation belts, this is probably the nearest desirable location that is acceptable for a long-lived infrared observatory. At this distance, the Sun dominates satellite heating and large areas of the sky are accessible for observations. Unfortunately, this orbit may be energetically the most expensive of those mentioned here due to the necessity for circularization.

2. *L2 Halo Orbit.* It may be possible to place a larger mass in an orbit ("halo") around the outer libration point than it is in HCO. At this location, about 1.5×10^6 km from earth, most of the sky is accessible and these libration point orbits have already been chosen for several spacecraft.

3. *Trans-Asteroid Orbit (TAO).* The ultimate celestial limitation to observations from the inner solar system at visual, near- and mid-infrared wavelengths is scattered light and thermal emission from zodiacal dust within the asteroid belt (Figs. 1 and 2). This material may be almost non-existent at solar distances beyond about 4 AU. Although some technical demands are great on a satellite so far from the Earth and Sun, the advantages of much reduced heating and reduction of the background limits by orders of magnitude make TAOs very exciting. Furthermore, using planetary gravity assists, it appears possible to launch a greater mass to this location than to any other that we are presently considering.

3.5. Thermal Model Testing & Verification Plans

It is too early in our study of EDISON to define the program for testing and verification in most areas, except to recognize that this remains a major issue. However, as our thermal design has progressed, it is possible to outline our plan for some assessments and verification.

As our design goal of a primary mirror temperature of 40–50 K is much lower than previous estimates that appear in the literature for hypothetical large space telescopes, we plan an extensive, early program of verification.

Our plan consists of three stages

1. *Extensive and multiple modeling.* We are undertaking analytic and numerical thermal modeling programmes to develop a wide and deep understanding of the limits to radiative cooling. Our results from different programmes will serve as a guide and verification for one another. No thermal modeling programme in the past has had as its goal the lowest temperature possible, which seems to be the primary reason why hypothetical future infrared telescopes have had baseline temperatures in the range 80–150 K. To begin development of EDISON, we must have confidence that the basic thermal structure of the facility is sound. Our first results appear in the paper by Hawarden *et al.* in this volume.

2. *Current radiatively-cooled telescopes in orbit.* With the thermal models for EDISON in hand, we will apply them to current space observatories. There are at present three radiatively-cooled telescopes in orbit that have roughly the same topology as proposed for EDISON and for which fairly reliable temperatures have been measured: IRAS, COBE, and HST. Indeed, as of this writing COBE has been operating as a radiatively-cooled observatory for over a year, collecting useful data after its cryogens boiled away. Similarly, all satellite subsystems on IRAS were operating *years* after its cryogens evaporated, except for the detectors which were designed to work only at a few degrees Kelvin. A major test of the thermal models will be whether they can duplicate the temperature structure of these three satellites.

3. *The thermal "wind tunnel."* Since the majority of parameters that govern thermal behavior scale well with size, our first tests of the EDISON design (and others under consideration) are planned to be scale models in environmental chambers. We are at present entering discussions over the availability of suitable environmental chambers.

4. Moderate-Aperture Radiatively-Cooled Infrared Telescopes

EDISON is currently being studied as the major international space mission to follow the current generation of purely cryogenically-cooled infrared facilities. Given that sensitivity of astronomical observations is a strong function of aperture under all conditions, we have proposed EDISON as possessing the largest aperture that we estimate can be constructed and launched with present or near-future technology. At the same time, we recognize the advantages of more modest satellites (that is, "Explorer-class" missions) in terms of technical feasibility, development time, and cost. Although such missions will be perforce less capable than EDISON as we have described it here, a $\sim$ 1 m infrared space telescope, radiatively-cooled to 30–40 K, will still have unsurpassed sensitivity in the near- and mid-infrared. It will also possess the other advantages of radiative cooling, including long life, lower cost for a given aperture, and the possibility of innovative optical and structural designs.

As part of our development of EDISON as a 3 m telescope, we are also investigating designs for radiatively-cooled infrared space observatories with apertures in the range of 1–2 m. Our work includes alternative optical and structural designs which indicate that temperatures are possible that are lower than the 40 K which we calculate for EDISON (Hawarden, Bally, and Thronson, 1992). Coupled with one of the more distant orbits that we are studying for EDISON (Section 3.4), a very exciting precursor to a large aperture radiatively-cooled telescope might be built and launched in the relatively near future.

Any capable space observatory will be expensive and we have been proposing from the beginning that EDISON be developed as an international consortium. The same would naturally be true of the more moderate mission that we outline here. Combining, say, the resources of an ESA Medium Mission and one of the larger NASA Explorer class programmes has the potential for a very capable 2 m class telescope in space. Additional resources from other nations may allow construction of a truly international infrared space observatory which approaches the full capabilities of EDISON.

5. The Scientific Case for EDISON

EDISON and possible smaller-aperture precursors are being proposed as the international infrared space observatories to follow the current first generation (purely cryogenic) missions. As such, they will carry highly capable photometric, spectroscopic, and imaging instruments which cover a wide wavelength range ($\sim$ 2–100 μm). This will allow a broad range of research programmes to be carried on for a decade or more by astronomers and space scientists from around the world. We cannot predict accurately what will be the most important research programmes in the early 21st century, but convention requires outlining hypothetical research programmes for the future.

Certain types of projects take advantage of the particular capabilities of EDISON: a large aperture operating at the celestial background limit at wavelengths shortward of about 30 μm is especially powerful to study small – or extremely distant – sources and structure and to work in confusion-limited environments. This is reflected in our list of possible research programmes.

5.1. Solar System Objects

Planets, satellites, comets, dust, and other solar system material have characteristic temperatures of a few hundred degrees Kelvin or less, making them ideal for observations in the region of maximum sensitivity for EDISON (Fig. 3). Purely cryogenic missions such as ISO and SIRTF are constrained from observing warm objects such as the Earth and its atmosphere or the Moon, as well as being limited to looking away from the inner solar system, thus eliminating infrared studies of, say, Venus, the solar corona, or comets during close solar passage. Radiatively-cooled telescopes, in contrast, do not suffer lifetime-limiting consequences of warm objects in the field of view. The effects of observing hot sources with EDISON are difficult to predict and is part of our thermal modeling of the spacecraft. We assume, however, that for several possible scientific projects, the most serious consequence is a warming of the telescope, which may limit subsequent observations to shorter wavelengths while the telescope cools down again. Alternatively, observations of the inner solar system or other warm regions may be undertaken early in the EDISON mission. It may be said, then, that EDISON will open the inner solar system to study from orbit in the infrared, including investigation of composition of [1] the atmospheres of Earth and Venus; [2] the material in comets during their closest passage to the Sun and, hence, during their most active period; [3] extra-solar system material on exceedingly small scales via lunar occultation; and [4] the solar corona.

Infrared observations of Mars have been outlined in these proceedings by Encrenaz, who points out that ISO and SIRTF should both contribute significantly to our understanding of the planet's atmosphere. EDISON will follow these observatories with higher spatial resolution investigation, especially if travel to a trans-asteroid orbit includes a close passage of the planet. Mineralogy from space of Mars will allow determinations of surface and dust storm compositions over large areas of the planet as a function of time.

Similarly for the Jovian planets, EDISON will provide new information on the atmospheres. For example, several major and minor molecular species have important transitions at wavelengths shortward of about 14 μm: CO, O_3, CO_2, H_2O, CH_3, and so on. At about 10 μm, EDISON at the diffraction limit will place about 1500 pixels across the disk of Jupiter from an orbit in the inner solar system. About 4 times this resolution will be possible in a trans-asteroid orbit. Furthermore, with sufficient spectral resolution, three-dimensional information about the atmospheric composition can be recovered.

Observations of comets will be a field of solar system study that will make special use of EDISON's advantages: [1] the observatory's reduced solar elongation requirement will allow greater coverage of the evolution of comets as they approach and recede from the sun; [2] the longer lifetime will increase the number of periodic and non-periodic comets (and other serendipitous events) that can be observed; and, of course, [3] the large aperture will allow observations of greater sensitivity and spatial information.

If EDISON is launched beyond the inner solar system or into an orbit that carries it out of the plane of the Earth's orbit, tomographic techniques may be used to construct a model of the zodiacal dust distribution. Somewhat similarly, if distant "Tritons" are discovered in the Kuiper belt (Stern, 1991), large aperture infrared telescopes may be the only tool capable of determining their sizes, mass, compositions, and, via astrometric observations, their distances.

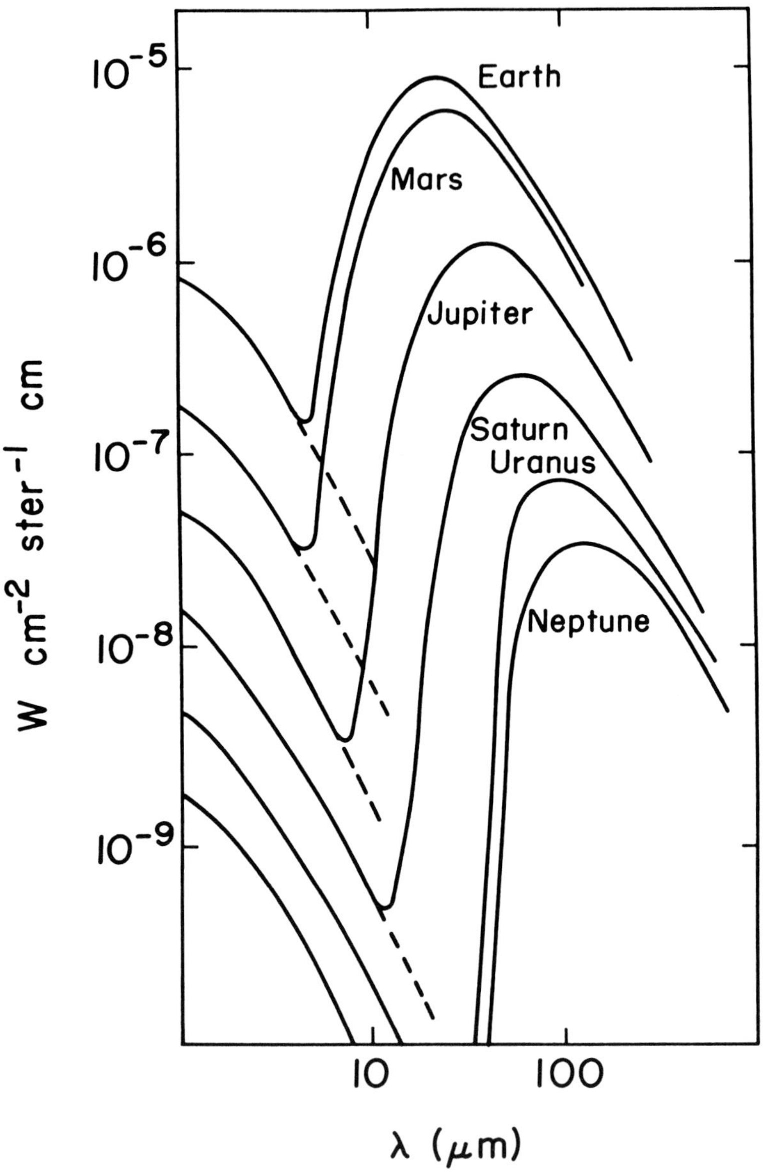

Fig. 3. Most of the scattered and emitted light from planets and sub-planetary objects occurs in the wavelength range of maximum sensitivity for EDISON. Investigation of solar system objects will include diagnostic spectroscopy of planetary atmospheres over a long period of time on arcsecond scale sizes or better and chemical composition of planetesimals.

5.2. Circumstellar Debris Disks or "Proto-planetary" Material

The discovery by IRAS of dust emission from nearby stars is exciting because of the role that this material may play in the early evolution of stars and their retinue, especially in relation to formation of planets. This material is impossible to study from ground-based telescopes, except for the few nearest objects, as observations from Earth's surface are either dominated by emission from the telescope and atmosphere or are at wavelengths completely obscured from the ground. EDISON, in contrast, will be limited only by the much weaker celestial emission and should be able to study "Vega disks" in detail throughout much of the Milky Way (Fig. 4). Equally exciting, EDISON will have the sensitivity and angular resolution to obtain broadband infrared photometry of individual main sequence stars in the LMC and SMC. If debris disks in the Clouds are as luminous as that around β Pic, spectrophotometry of the dust emission can be obtained in an integration time of 1–2 days over a large area in our companion galaxies. This allows an investigation into debris disks in systems with much lower heavy element abundances than the Milky Way.

The large aperture of EDISON will allow us to study spatial structure in the circumstellar disks: the diffraction limit of a 3 m telescope at 10 μm is $1''$ or 10 AU at a distance of 10 pc. Advanced image analysis techniques will improve this resolution by a factor of 2 or 3. Thus, structure in the disks on scale sizes comparable to the separation between planets in our solar system can be studied. For example, the dusty material around Vega is estimated to lie between 50 and 100 AU from this star. Thus, in the mid-infrared EDISON will have more than 300 pixels across the region occupied by the disk. Just as important, EDISON's spectroscopic capability will be used to determine the chemical composition of the dust and the variation of this composition with position.

5.3. Brown Dwarfs and Other Substellar Objects

Objects more massive than planets, but not sufficiently heavy to shine via hydrogen fusion are among the most intriguing objects yet to be studied. They may be a major class of celestial object, possibly contributing to the "dark matter." Moreover, they are the link between planets and stars. Brown dwarfs are cool, however, with estimated surface temperatures in the range of $\sim$ 1000–2000 K about a billion years after birth, dropping further to a few hundred degrees Kelvin for the oldest brown dwarfs. A large fraction of their light is therefore emitted longward of 3 μm, which has kept them from being easily detected. If ever found, they will be difficult to study in detail as much of their emission will be either masked by terrestrial atmospheric absorption or overwhelmed by thermal emission from warm telescopes.

EDISON will excel at studying brown dwarfs and related objects. We assume that by the time that the observatory is launched, brown dwarfs will have been unambiguously discovered and, perhaps, found to be abundant. If not, a celestial-background-limited large aperture infrared telescope with large arrays will be exactly the facility necessary to search deeply for these elusive objects. In the event that brown dwarfs will have been identified in the coming decade, we believe that moderate- and high-resolution spectroscopy will be the most critical tool of investigation. Broadband photometry will still be important, of course, for example in searching for excess emission from faint (planetary?) companions or circumdwarf material. Brown dwarfs may be discovered more readily as companions in stellar systems, which will require higher angular resolution to separate components in the closer systems. In addition, it may be the case that astrometry with EDISON will be necessary to determine distances to brown dwarfs, as well as other cool objects. This will be an especially attractive goal if the satellite is launched into a trans-asteroid orbit.

However, just as with stars, spectroscopy of brown dwarf atmospheres will reveal key characteristics of these objects. Mass, age, luminosity, and chemical composition – the traditional parameters in stellar evolution – significantly alter dwarf photospheric spectra. For example, variations in the carbon-to-oxygen abundance ratio, as well as the presence of grains, will produce very distinct features even in low-resolution spectra. Furthermore, the broadband spectral energy distribution is a strong function of age and mass. Happily for possible planetary detection around brown dwarfs, predicted emergent spectra for the failed star are very different from that for giant planets.

5.4. Evolved Stars and Their Surroundings

Evolved stars return processed nuclear material to the interstellar medium, may be the primary factory for grain formation, and are versatile environments for the creation of complex molecules. They are also among the most luminous individual stars at infrared wavelengths, capable of being studied at vast distances, which will allow investigation into the characteristics of these objects in external galaxies. Work in such crowded fields will tax EDISON's angular resolving power, but the rewards of a complete understanding of highly evolved stars will be worth this effort. Characteristic temperatures of the material surrounding dying stars are in the range of 50–1000 K, which means that key gaseous and dust diagnostic features emit strongly in the wavelength range most accessible to EDISON.

For example, the 43 μm water ice band has now been observed in a number of oxygen-rich evolved stars and it has been used to constrain the temperature of the region of formation, as well as the cloud conditions that favor amorphous or crystalline grains. Similarly, it is probable that outflowing material around evolved objects is the source of polycyclic aromatic hydrocarbons (PAHs) or other very large molecules which may produce the infrared "unidentified features." These bands are often the most intense spectral features in star-forming galaxies, thus they play a major role in regulating the transfer of radiation within the ISM. Their structure, composition, and mechanism of formation is critical for a complete understanding of the material between stars. EDISON can be used to obtain the necessary low-resolution spectra of a very large number of highly evolved objects, with the goal of determining the evolution of the grains within the dusty envelopes. Because of the luminosity of these objects, they will be able to be observed easily, for example, in both Magellanic Clouds, thus significantly increasing the environments in which circumstellar grain evolution can be investigated.

Low-spectral-resolution spectrophotometry can be used to classify circumstellar shells in galaxies with a wide range of gross properties, thus allowing investigation into mass return, evolution of the ISM, and the role of chemical abundances as a function of, say, galaxian metallicities. This may be particularly important in the study of star-forming metal-poor dwarfs as analogs to primeval galaxies.

Some similar considerations apply to planetary nebulae and their immediate precursors, where several highly-ionized atomic species have important lines in the infrared, including hydrogen and helium, of course, but also neon, nitrogen, and chlorine. Broad dust features of SiC at 11.2 μm and, perhaps, MgS at about 35 μm, along with several of the shorter-wavelength "unidentified features" promise that both the gas and dust phase of the circumstellar material can be studied in detail in these objects.

5.5. THE INTERSTELLAR MEDIUM IN GALAXIES

The interstellar medium (ISM), both dust and gas, is cool, with characteristic temperatures in the range of about 5–200 K, which means that study of the structure of this material, its evolution and composition, and the birth and death of stars must be undertaken in the infrared and sub-millimeter regime. The spectral coverage of EDISON will include the near-infrared, with its diagnostic lines of atomic and molecular hydrogen, to the far-infrared, with emission from dominant atomic cooling lines and higher rotational levels of important molecules. Between these two wavelength extremes, EDISON will be used for study of the mid-infrared ground-state molecular hydrogen lines and will be a major probe of the most abundant molecule in interstellar space.

The gaseous lines are superimposed upon the broad thermal dust emission and infrared solid state features, including emission and absorption bands of silicates, H_2O, CH_3OH, CO, CO_2, and several strong unidentified features. The maximum in the flux distribution from grains in the ISM often occurs in the 50–100 μm wavelength regime, which means that EDISON will be able to map the distribution of maximum energy emission in star-forming regions with an angular resolution of $5''$–$10''$. This is about the scale size of spiral arms and the interarm region in fairly nearby galaxies. Characteristically, the near- and mid-infrared solid state dust features are the strongest spectral signatures in star-forming objects. Although many mysteries remain about these broad features, by the time that EDISON is launched, we expect that they will be available as primary diagnostics of the composition and abundance of the ISM.

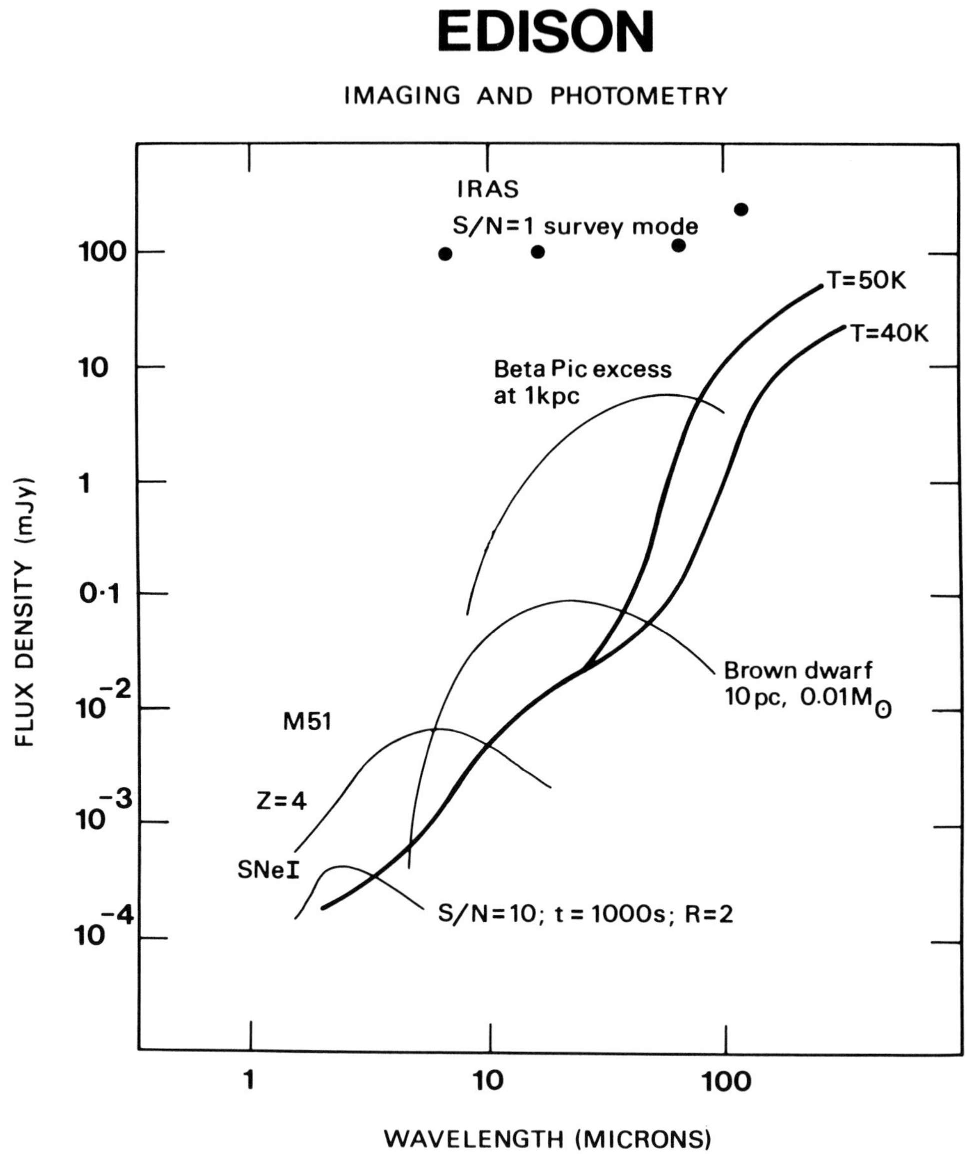

Fig. 4. Sensitivity of broadband EDISON photometric and imaging systems under the assumptions of diffraction-limited point source observations at the background limit, $\lambda/\Delta\lambda = 2$, and an integration time of 1000 seconds (17 minutes) to reach a signal-to-noise ratio of 10. Two equilibrium temperatures for the primary are shown, 40 K and 50 K. Also represented are the broadband spectral energy distributions of four interesting celestial objects: the circumstellar "debris" disk around β Pic as it would appear at 1 kpc, a model brown dwarf, and the Whirlpool Galaxy and a Type I supernova (maximum light), both at $z = 4$.

5.6. THE BIRTH OF ELEMENTS: HIGH-z SUPERNOVAE AND THE EVOLUTION OF THE ISM

There are only a handful of truly key events in the history of the universe: its birth, the time of recombination, the formation of the galaxies, and the beginning of enrichment of the ISM with heavy elements created in the cores of massive stars. This last event fundamentally altered the mechanisms of star formation and set the universe on a course that led to planetary formation and the appearance of life. Indeed, so important is an enriched ISM, that it is fair to call its appearance as the dawn of the modern universe. However, at present we barely have a fair idea of how the galaxies themselves formed, to say nothing of when and how they began to alter their ISM.

Observations of the gas phase of the early universe will no doubt be important and EDISON will contribute significantly to study of, for example, the Balmer, Paschen, and Brackett lines in high-z objects. These lines are strong and, even redshifted, remain in a dark portion of the celestial background. Low-excitation H_2 lines, such as those at 28 μm and 17 μm, will also allow us to study the conditions associated with early generations of star formation. Similarly, there are photospheric features, such as the red TiO and near-infrared CO and H_2O lines, especially strong in red giants and supergiants, which can be used as tracers of advanced stages of stellar evolution, as well as isotopic photospheric abundances in high-z galaxies. Spectroscopic observations of these features may also allow a determination of the large-scale stellar motions in primeval objects, which will open the study of the dynamics of galaxies in the very early universe. Detection of such strong features is not difficult: Figure 4 demonstrates the ease of broadband detection of even normal galaxies such as M51 at cosmological distances. Even dwarf galaxies, which may have been the building blocks of the early universe, can be imaged by EDISON at vast distances.

Given the progress of the last few years in understanding the infrared solid state features (in some cases referred to as "unidentified infrared features," but here we also include the mid-infrared "silicate" and 3.1 μm ice features), we expect that by the time that EDISON is launched, the infrared bands will be major diagnostic features of the ISM. This should be a major goal of current theoretical and laboratory astrophysics as these features can often be the strongest spectral signature in star-forming galaxies in the local universe. Based on current understanding of these features, a major program for EDISON will be the study of the excitation, composition, and physical structure of first-generation interstellar grains in young galaxies. Dust grains constitute a major component by mass of the heavy elements between stars, so that they are intrinsically interesting. They also are likely to dominate the cooling and chemical evolution of the ISM, critical physical processes in the early history of the universe.

Essential to these observations, as well as so many others, will be the sub-arcsecond angular resolution of EDISON throughout the near- and much of the mid-infrared. It will be critical to be able to study the spectral features as a function of position within high-z and primeval galaxies. For example, imaging of the redshifted atomic and molecular hydrogen lines in primeval galaxies will allow us to study the birth of the first generations of stars. Not only is a large aperture necessary for high-resolution imaging, but important scientific investigations might be seriously compromised in the confusion-limited environment of the distant universe. This has been one of the reasons why the EDISON programme has identified large aperture as one of the paramount programme goals.

Figure 4 also shows that EDISON will have no difficulty in determining the spectral energy distribution of Type I supernovae at cosmological distances, if they exist. There will, therefore, be no difficulty in using these objects as distance indicators throughout the universe. Young Type II SNe are a few magnitudes fainter than Type I, with more unpredictable maximum light, but these objects should be detectable at high z with an integration time of a few hours. With sub-arcsecond resolution in the near-infrared, EDISON can map Type II SNe in extremely distant objects, thus directly determining the rate and location of high-mass star formation in primeval galaxies. Finally, there may be classes of SNe that existed in the early universe, but are now extinct. An inventory of SNe types at vast distances will be revealing of the changing nature of high-mass star formation and the creation of the elements.

6. Summary

EDISON is under preliminary study as the international large-aperture, long-lived successor to this decade's generation of cryogenic infrared space observatories. Our current design is for a Cassegrain telescope roughly 3 m in diameter which will cool via radiation to $T \approx 40$–50 K. As a consequence, the facility will be celestial-background-limited throughout the near- and mid-infrared and will possess low background even in the far-infrared. The large aperture will mean a major increase in point-source sensitivity and angular resolving power, as well as much improved performance in the confusion-limited environments of crowded star fields and the distant universe. Our consideration of more modest-sized precursors is also outlined, including the possibility of combining the resources of an ESA Medium Mission with a NASA Explorer. We briefly point out the possibility of even colder telescopes via radiative cooling using innovative structural and optical designs which we are also investigating.

There are an enormous number of scientific projects that will be proposed for EDISON, only a few of which we described here. Some highlights include composition of planetary atmospheres and comets, analysis of brown dwarf photospheres, circumstellar and (perhaps) planetary material around other stars, star formation and the birth of the heavy elements in the distant universe, and the structure of very young galaxies.

Acknowledgements

We appreciate discussions with other members of the EDISON working group, especially J. Bally and C. Telesco, as well as other authors in this volume and participants in this year's workshops in Edinburgh and Boulder on next generation infrared observatories. Alan Stern has been one of the most effective advocates for a trans-asteroid orbit and Mike Werner discussed with us the estimated performance of SIRTF. Preliminary study of large-aperture, radiatively-cooled infrared space telescopes has been supported in part by C. H. Thronson, the Royal Observatory, Edinburgh, and NASA's Astrotech 21 program of the Advanced Projects Office. We appreciate the continuing advice and support of M. Kaplan and L. Caroff at NASA Headquarters and senior scientists at the Royal Observatory, Edinburgh, including J. Bell Burnell, T. J. Lee, M. Longair, P. Murdin, and R. Stobie.

References

Frisk, U. O.: 1991, in Y. Kondo, ed(s)., *Observatories in Earth Orbit and Beyond*, Kluwer Academic Publishers, 223

Gautier, T.N., Boulanger, F., Pérault, M., and Puget, J.L.: 1992, *AJ*, **103**, in press.

Hawarden, T. G., Bally, J. and Thronson, H. A.: 1992, *in preparation* ,

Hawarden, T. G., Cummings, R., Telesco, C. M. and Thronson, H. A.: 1992, *Space Science Reviews* , this volume

Helou, G. and Beichman, C.: 1991, in EDITOR, ed(s)., *From Ground-based to Space Borne Sub-Millimeter Astronomy*, 29th Liege International Astrophysical Colloquium,

Humphries, C. M., Nevo, Y., Ettedgui-Atad, E., and Harris, J. W.: 1992, *Space Science Reviews* , this volume

Lester, D.: 1991, in A. Fillipenko, ed(s)., *Symposium on Robotic Telescopes*, ASP Symposium Series, in press

Orlowska, A. H., Bradshaw, T. W. and Hieatt, J.: 1990, *Cryogenics* **30**, 246

Phillips, T.: 1991, in Y. Kondo, ed(s)., *Observatories in Earth Orbit and Beyond*, Kluwer Academic Publishers, 231

Stern, A.: 1991, *Icarus* **90**, 271

EDISON MISSION: LAUNCH VEHICLE AND SPACECRAFT CONSIDERATIONS

R.F. TURNER and W.J. CURTIS
Space Science Department, Rutherford Appleton Laboratory, Chilton, Didcot, Oxon OX11 0QX, U.K.

1. Introduction

The basic concept for the EDISON programme is a long-lifetime (10 year minimum) infrared astronomy mission. Coupled with this is the requirement for extended, uninterrupted observing periods. These two brief statements drive two major areas (i) instrument and detector cooling and (ii) delivery vehicle. As with all spacecraft designs, each system parameter interacts with another system parameter. Given the spacecraft guideline specification this paper addresses a number of these areas to create a forum for further discussion. (The scientific basis of the proposed mission will not be discussed here, since details appear elsewhere in this issue.)

2. Spacecraft Guideline Specification

For the purposes of this paper the following spacecraft guideline specification has been assumed:

- orbit – geostationary transfer orbit, high eccentric or Lagrange,
- mass – 4000 kg,
- observatory length – 6.6 m (comprising 4 m telescope, 1.6 m focal plane instruments, 1 m sun shield),
- observatory overall diameter – 4.6 m (see launcher survey),
- service bus – 1.5 m length, 750 kg dry mass,
- propellant mass – up to 350 kg propellant depending upon orbit (N.B. assume that only 20% is needed for station keeping and attitude control system),
- solar array mass – 75 kg,
- spacecraft overall length – 9.1 m,
- earliest launch date – 2005.

3. Launcher Survey

A survey has been conducted of the approximate performance parameters for current and future launchers. The survey excluded Shuttle and Buran/Energia launchers

Space Science Reviews **61:** 171–185, 1992.

TABLE I
Launcher data (single-passenger spacecraft only)

	Payload lift mass kg						
Vehicle	LEO	Sun Synch	GTO	GEO	Fairing dia (m)	Fairing length (m)	Comment
Ariane 40	4600	2700	1900		3.65	8.6	or 9.6m or 11.1 m long
Ariane 441	7000	6000	4400		3.65	8.6	or 9.6m or 11.1m long
Ariane 5	18000	12000	6920		5.4	12	or 18m long
Pegasus	408	272			1.3		
Atlas 2 Centaur	8610		3630		3.7	7.8	2100kg Escape
Zenit	15000		5900	2400	3.9		
Proton	21000		4600	2200	3.3	6.2	
CZ2 E	8000		2500		4.2		
Scout 2	520				1.4	2	Length estimated
Titan Commercial	14700		5000		3.65	10	
Titan 4 Centaur	17500	14500	4500	5500	4.6	22	
NASDA H1	2250		1100	550	2.4	7.9	Restricted launches
NASDA H2	9400	4500	4000	2000	4.1	12	Restricted launches
DELTA 2	5000	3800	1800		3.1	7.9	or 2.8 x 8.5m

Notes:
1. Where available the Fairing diameters quoted include dynamic clearance.
2. Fairing lengths should be treated with caution. Many vehicles have a range of fairing lengths. The shortest is quoted here. Longer ones may reduce payload available mass. Fairing lengths include non-parallel sections.
3. This data is taken from published material and may contain inaccuracies.

because (i) by 2005 the Shuttles will be old and will be reserved for Spacestation resupply duties and (ii) the Soviet economy is not expected to sustain development of large scale systems. A further assumption is that any new launchers that are developed will be intended for heavy lift to low earth orbit and will therefore not be applicable to the EDISON mission.

The launcher survey data are listed in Table 1. The lift capabilities to low earth, sun-synchronous and geostationary transfer orbits (GTO) are shown graphically in Figure 1. The launcher fairing dimensions are similarly shown in Figure 2. Caveats abound about the accuracy which may be assumed for the data and particular caution should be exercised in accepting the fairing diameters. The proper criterion is the available dynamic envelope which takes account of the deflections of the rocket nose cone during launch. The figure quoted for the TITAN 5m diameter fairing is an assumed dynamic envelope of 4.6m.

Examination of Figures 1 and 2 leads to the following conclusions regarding

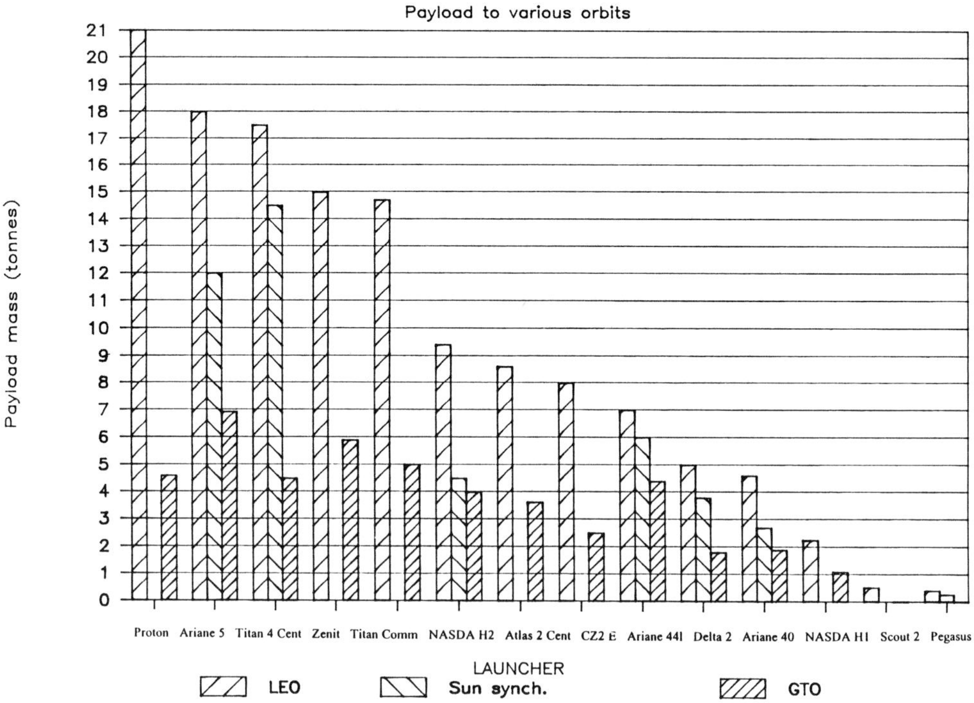

Fig. 1. Launcher performance.

suitable launchers:

1. Complying with all criteria in Section 2 above:-
   Ariane 5, Titan-Centaur
2. Complying if diameter is relaxed to 3.65m:-
   Ariane 5, Titan-Centaur, Titan-Commercial, Ariane 44L, Zenit, NASDA H2
3. Complying if diameter is relaxed to 3.3m:-
   Ariane 5, Titan-Centaur, Titan-Commercial, Ariane 44L, Zenit, Proton, NASDA H2

All of the launchers will need to be re-assessed when the orbit is finalised to ensure that their stage assemblies can be optimised to achieve the goal (e.g. some vehicles are optimised for GTO performance). Additionally the launch sites of some vehicles will restrict orbit inclinations (e.g. with the Atlas 2 there are no facilities for polar or sun-synchronous launches).

If a Lagrange orbit is desired then it should be noted that it is planned to place the 1850 kg SOHO spacecraft in a halo orbit at the L1 point by means of an Atlas 2 rocket. Since the Atlas has a GTO capability of 3630 kg then *pro rata* the Titan-Centaur should be capable of placing 2293 kg into a halo orbit at the L1 point. (Interestingly the Titan-Commercial may possibly do better at 2550 kg.)

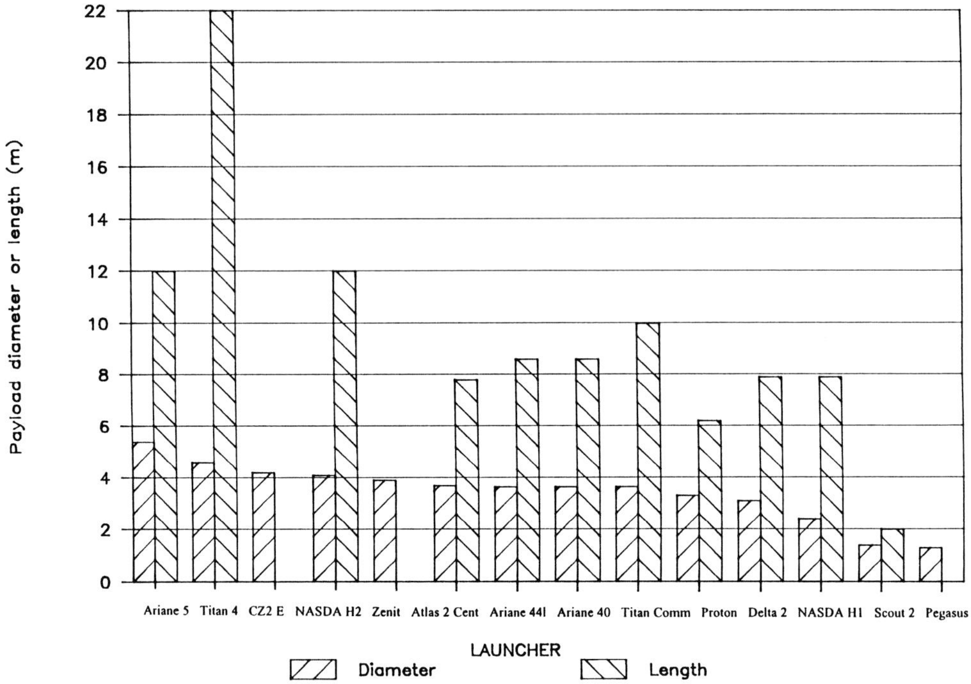

Fig. 2. Payload fairing dimensions.

In summary all this suggests that the study should continue with the Titan-Centaur as the baseline, since the Ariane 5 launcher is not yet proven.

4. Cooling System

The 10 year lifetime specified exceeds by a factor 5 to 8 the typical lifetime of a cryogenically cooled mission such as IRAS or ISO. As a similar thermal performance is still required this implies that a new approach is needed.

Cryogenic cooling. A vastly improved system design would be needed, but the required improvement is unlikely to be achieved. It should be possible to assist the cryostat performance by reducing the external temperature and hence the thermal load, but experience on IRAS and ISO indicates that the improvements expected can only be marginal.

Mechanical cooling. Much work has already been conducted in this field with the UK Stirling cycle cooler (single stage) to be launched this year on ERS1 and UARS. These will be the first such devices in orbit. Currently the ground life accumulation on a single unit stands at about 2 years. Thus the maturity and lifetime of this approach still has to be demonstrated. Nevertheless many earth observation programmes are basing their missions on this technique and we can expect to see steady developments over the next decade. Multistage coolers for detectors could be combined with single stage units cooling the enclosures. Typical hybrid arrangements, as have been suggested for the SAFIRE radiometer proposed for a Polar platform, are described in the Appendix.

The approach most likely to bear fruit is a passively cooled spacecraft (in which case one face always points to deep space) with mechanically cooled detectors. However, other refrigerant cycle systems will need to be evaluated as a prime solution or as part of a hybrid system for intermediate cooling.

5. Spacecraft Systems

5.1. Coolers

These will be mission-critical. In view of their complex nature and lack of maturity, duplication or even triplication is necessary. One implication is that the one way flow of "cooling" becomes a requirement so that the redundant, non-operating system does not itself get cooled. A possible solution to this would be to use thermal diodes. (These are devices which would be placed in the thermal link to the cooler and which use differential expansion of their component parts to break the link if they rise above a set temperature.)

5.2. Spacecraft Power

Broad guesses at the power consumption are as follows:

Service bus consumption	500 W
Instruments/telescope	250 W
Cooling systems	200 W
TOTAL	950 W

5.3. Array Sizing

Assume a 30% degradation of the solar array in a 10 year period due to cover glass darkening and radiation damage. Thus to maintain the power requirement to the end of life requires 1235 W initially. Allowing 70 W m^{-2} gives an array size of 17.6 m^2. With an array height of 6m this requires a width of 3m. As this is the same order as the diameter of the spacecraft it may be possible to dispense with or at least limit the use of a deployable array.

TABLE II
Communication Characteristics

Telecommand	2 KBPS	
	Probability of frame rejection $< 10^{-6}$	
	Probability of false command $< 10^{-8}$	
Telemetry	1.2 KBPS/47.2 KBPS/211 KBPS data rates	
	Reed Solomon plus convolutional coding	
	Bit error rate: 10^{-7}	
	Source packet delivery probability: 99.999%	
Tracking	Deep space tracking system	
	Ranging carrier/noise density > 10 dBHz	
	Ranging signal/noise density > 1 dBHz (IF filter implemented in HPA)	
Receiver	Noise Figure < 5 dB	
Transmitter	8.9 W RF output power	
Antenna	2 LGAs	±108.5
		> −9.3 dB Reception right circular polarization
		> −7.0 dB transmission
	HGA	0.8 M Dish
		> 14.6 dB Reception gain/right & left circular polarization
		> 18.2 dB transmission
		±32 displacement by pointing mechanism

Note:
Example derived from SOHO mission – (Matra study)

5.4. Spacecraft Bus

Matra have recently completed a study for the bus section of the SOHO spacecraft which also uses the Lagrange orbit. For the first approximation to an EDISON bus it would be wise to base it on this study. The specification of the data and telemetry details of the bus are listed in Table II.

5.5. Attitude Control System

Unless a low earth orbit is chosen the attitude control will be by reaction wheels off-loaded against a gas jet system. It is worth noting that IUE uses only 0.5 kg of gas per year as a result of careful target planning.

6. Spacecraft Configuration

The spacecraft configuration will depend extensively on the orbit chosen. The different stable attitudes of the spacecraft as a function of type of orbit are shown in Figure 3. One possible configuration is shown in Figure 4. The direction away from the Sun is labelled Q. For the sun synchronous orbit (Figure 3a) Q always faces deep space and the telescope views the earth once per orbit, which would be unacceptable for a cryogenic telescope such as this. For the geostationary earth orbit (Figure 3b), and for any other equatorial orbit, the telescope always views deep space and Q faces the earth once per orbit. For the Lagrange halo orbit (Figure 3c) Q always faces the earth and the telescope always views deep space.

In all configurations an important design point is to avoid the possibility of the telescope externals viewing the rear surface of the array (which does get hot). It is not shown in Figure 4 but the array should be deployed away from the telescope with a significant gap between it and the heat shield (which in turn has an established gap between itself and the telescope). This will minimise the radiative transfer of heat from the back of the solar array to the telescope.

7. Recommendations for Future Study

7.1. LAUNCHER

Adopt TITAN-CENTAUR as baseline. Investigate the capability for the orbit required. Confirm dynamic envelope and any centre of gravity problems.

7.2. ORBIT

The spacecraft needs to maintain a specific orientation with respect to the Earth-Sun line such that one face always views deep space (for the thermal dump) and another face always views the sun (for the power and defined heating direction). The heating effect of the earth view should not be discounted. Such restrictions will have a profound effect upon the telescope pointing. The orbit selection may require that the path of the earth around the sun will dictate observation windows over the course of a year. The best options would be a high earth orbit (say 70 to 100,000 km) to minimise the thermal effect of the earth or the Lagrange halo orbit (which would eliminate this effect altogether).

7.3.

The spacecraft service bus should be modelled upon the SOHO type (though not the solar array).

7.4.

The solar shield and array should be designed to provide a two-stage shield as shown in Figure 4.

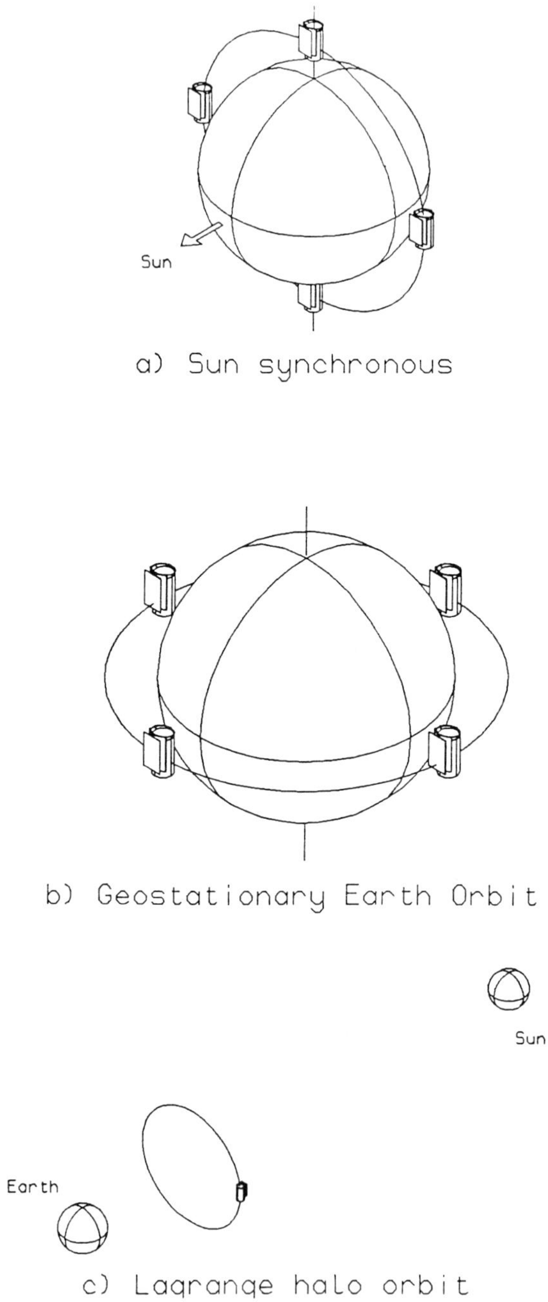

Fig. 3. Spacecraft configuration, as a function of different orbits.

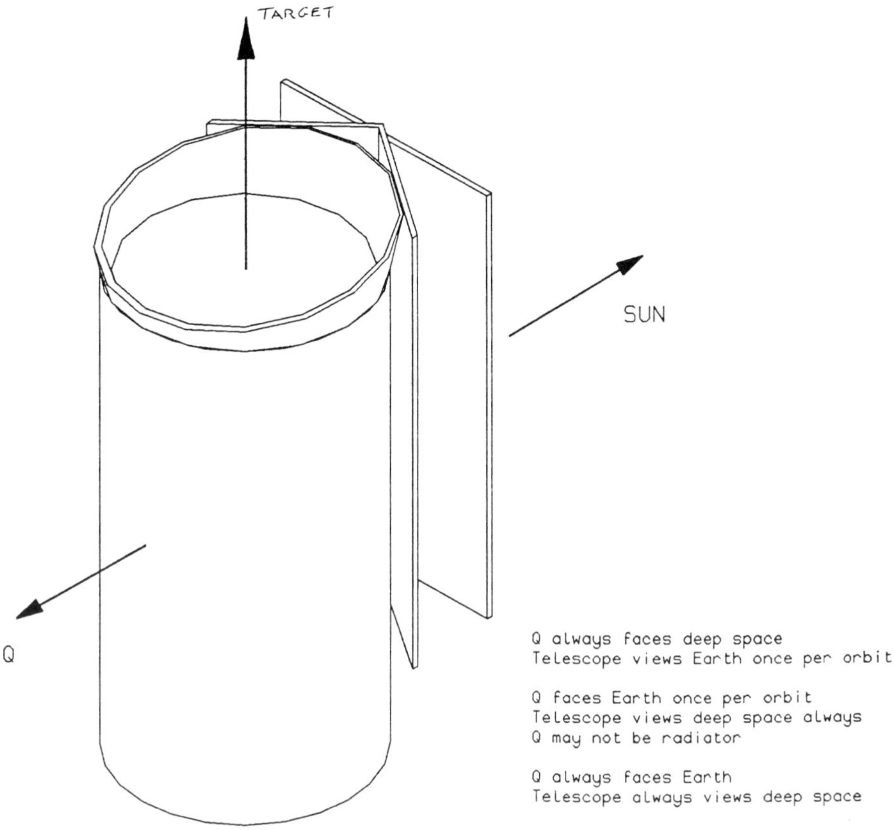

Fig. 4. Suggested arrangement of spacecraft construction showing solar array and sunshield. In practice the solar array would be displaced away from the telescope to minimise any radiative heating of the telescope by backside of solar array.

7.5.

Encourage the development of a thermal diode system to allow the use of redundant mechanical coolers.

7.6.

Stirling cycle coolers are not the only solution. Other refrigerant cycles should be evaluated as a prime or hybrid solution.

7.7.

Greenspace! It may be law by the year 2005 that objects in space need to be placed into an orbit which will decay in a short period. Should we start by embodying a fuel reserve which could accomplish this?

Appendix

(Text quoted from the unpublished SAFIRE Phase A Study.)

Hybrid cooling arrangements for SAFIRE

A. Far IR Detector Cooling

The far-IR Ge:Ga detectors require cooling to below 4.5K for a period in excess of 5 years. A preliminary study of this requirement has been conducted, and the two options which present the best solutions are:

1. A hybrid superfluid helium dewar with active mechanical cooling of its shields. Stirling cycle coolers lower the temperature of the radiation shields and significantly reduce the heat load on the liquid helium, extending its life.
2. An all closed-cycle system which uses closed-cycle coolers to provide a pre-cooling stage for a 4.5K Joule-Thomson (J-T) system. The J-T stage uses the same type of long-life compressors used in the Stirling cycle.

Figure A1 is a schematic block diagram of the cryogenic subsystems. Both our baseline and J-T option approaches for long-life cryogenic cooling depend upon cooling to 30K by mechanical Stirling cycle coolers.

The hybrid helium dewar is the SAFIRE baseline choice for two reasons. First, a considerable design and experience base has been gained with the IRAS and COBE space qualified dewars. Also a detailed study has recently been completed for the XRS (AXAF) experiment [(AXAF XRS Cryogenic Subsystem Definition Study, Final Technical Report Contract NAS5-30186, Dec. 1987, Addendum Apr. 1988)] which demonstrates the performance improvements of using active (mechanical) cooling to extend liquid cryogen lifetimes in a space dewar. Second, this option requires less power than the closed cycle cooler (140 vs 342 W). However, it does occupy a larger volume and has a greater mass (23 kg).

Both of these options rely upon the single-stage Stirling cooler technology which has been developed at the RAL with ESA and UK funding. A development model two-stage Stirling cycle cooler is currently operating at RAL, and achieves a cooling power of 100 mW at 25K and 200 mW at 30K. These cooler capabilities are compatible with the cryogenic requirements for either option. Thermal power dissipation is a major consideration in the cryogenic cooler design. It appears at this stage of our work that the J-T cooling option increases the power requirement above the level that could be reasonably radiated to space, and therefore, would require an active cooling interface with the parent spacecraft. The possibility of improved cooler performance (lower power requirement) or an improved heat rejection capability will influence the final choice. For these reasons, both options will be studied further in the definition phase.

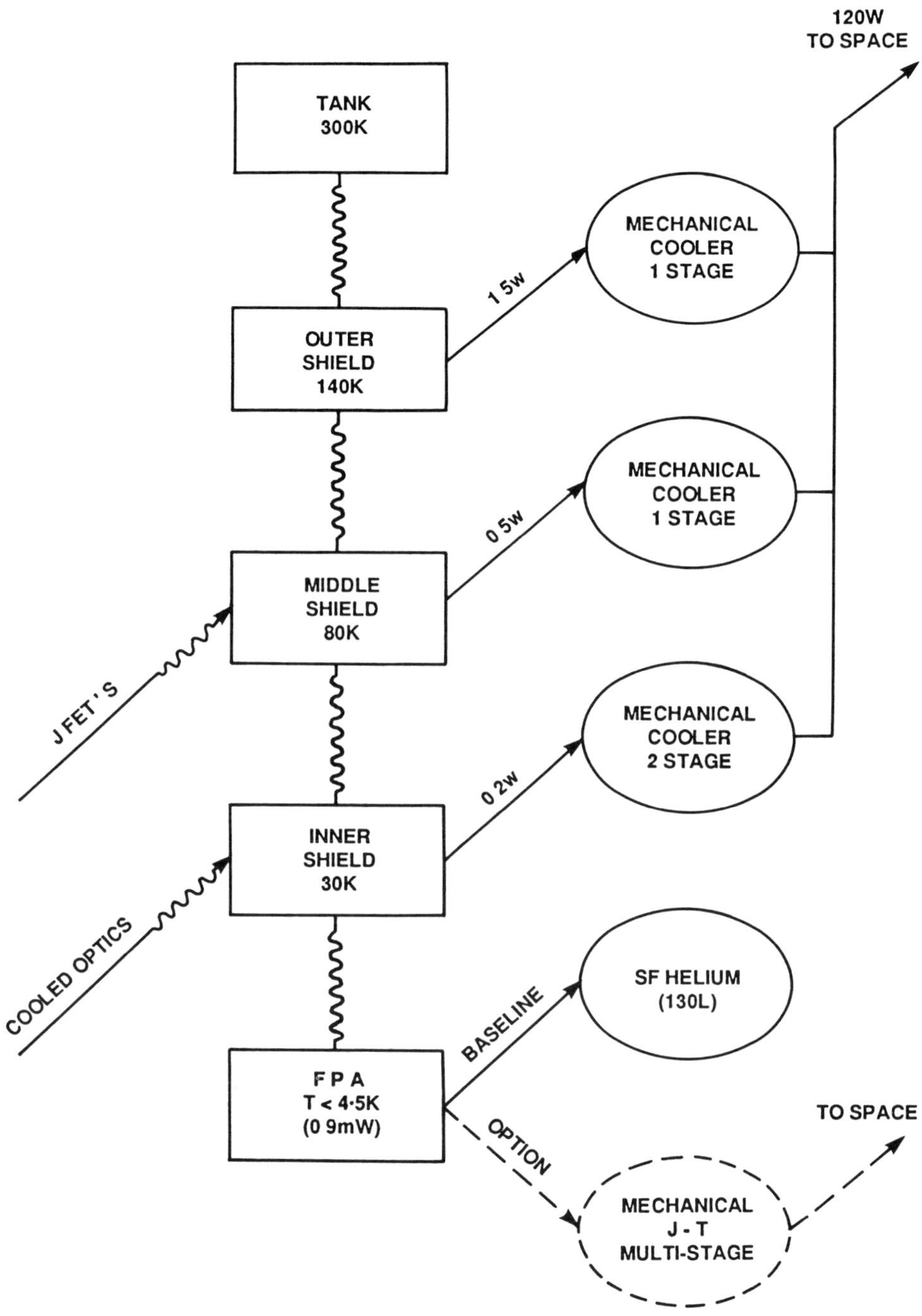

Fig. A1. Schematic diagram of the far-IR detector cooling system. The base-line design uses Stirling cycle coolers to maintain CODM shield temperatures of 140K, 80K, and 30K and employs liquid helium to cool the detector array to < 4.5K.

TABLE A I

4K stage heat load budget

Signal	< 0.01 mW
Radiation	0.5
Sensor Dissipation	0.05
Support Straps to FPA	0.3
Wiring	0.05
TOTAL	0.9 mW

B. Long-Life Cryogenic Cooling

B.1. Hybrid Dewar Cooler Preliminary Analysis

In the hybrid dewar baseline design, the focal plane assembly is supported with a secondary tensioned strap attachment to the 30K subsystem. Tensioned graphite-epoxy straps provide the best combination of low conductivity and high strength for structural supports below 30K. The JFET transimpedance amplifiers will be structurally mounted to the 30K optical bench adjacent to the FPA, but thermally tied to the 80K stage. The detector-to-JFET interconnecting leads will consist of vacuum deposited manganin wires on mylar ribbon similar to the design of IRAS. The total leads to the FPA will consist of 50 for sensors, 12 for temperature sensing, and 4 for cryogenic blackbody sources. Table A1 summarises the heat to the 4K stage.

Based on this CODM data, a preliminary dewar sizing has been determined using the Ball Aerospace Cryogenic/Thermal Analyzer Program, and the results are given in Table A2. The amount of liquid helium required for a 5-year mission as a function of focal plane heat load is shown in Figure A2. This analysis shows the dramatic increase in on-orbit liquid helium cooling lifetimes which are enabled by the incorporation of mechanical cooling to temperatures of 30K.

Ball Corporation design data was also used to estimate the cooling loads for the actively cooled shields (ACS) in the SAFIRE dewar design. These values rely on the COBE flight hardware validated model which does not fully represent the performance of the latest better optimized tensioned strap support system. The shield cooling loads are as follows:

OACS at 140K	1.5 W
MACS at 80K	0.5
IACS at 30K	0.2

These values are conservative when scaled to compare to actual test results obtained for the COBE dewar. The COBE dewar has about six times the shield area and for a room temperature outer vacuum shell temperature, the equilibrium

TABLE A II
Dewar design parameters

Experiment heat load	0.9 mW
Lifetime	> 5 years
Volume of liquid helium	130 1
Mass of liquid He	17 kg
Total mass of full dewar (excludes cold optics/FPA)	63 kg
Support strap conduction	1.1 mW
Radiation through MLI	0.5 mW
Inner shield (30K) heat load	0.2 W
Middle shield (80K) heat load	0.5 W
Outer shield (140K) heat load	1.5 W
Total compressor power	140 W

shield temperatures were 165, 85, and 44K; with measured loads of 2.46, 1.30, and 1.27W respectively. Adjusted for the area difference, this COBE test data predicts 0.6, 0.22, and 0.21W for SAFIRE shield loads. These cooling requirements are compatible with present performance of UK-developed single- and two-stage split-cycle Stirling coolers. The three shield design minimizes total compressor requirements but may impose some mass penalty. This optimization will be studied further in Phase B.

The use of a two-stage Stirling cycle cooler to attain active cooling to 30K is critical to achieving the desired lifetime within the size and power constraints of this experiment. In terms of technology development, this requires a demonstration of a flight qualifiable two-stage cooler in the ESA-funded effort at Rutherford Appleton Laboratory. This is a realistic expectation; a development model of a two-stage Stirling is presently achieving 200 mW at 30K, and its main components, such as compressors, are similar to those of the already flight qualified single-stage coolers marketed by British Aerospace Engineering.

B.2. Closed Cycle J-T Cooler

B.2.1. Design Principles and Requirements

A study has been conducted on closed cycle cooling options for future space missions (Aerospatiale, 1984). The results of this study indicated that for low cooling power at 4K, a Joule-Thomson cycle pre-cooled by a two-stage Stirling cycle was the preferred option.

The layout for the 4K system is shown in Figure A3. The basic elements in this design are pre-cooler, J-T valve, isothermalizing region, J-T heat exchanger, gas cleaners, second-stage heat exchanger, and first-stage heat exchanger. The J-T valve will be a fixed orifice device constructed from a capillary tube, although porous

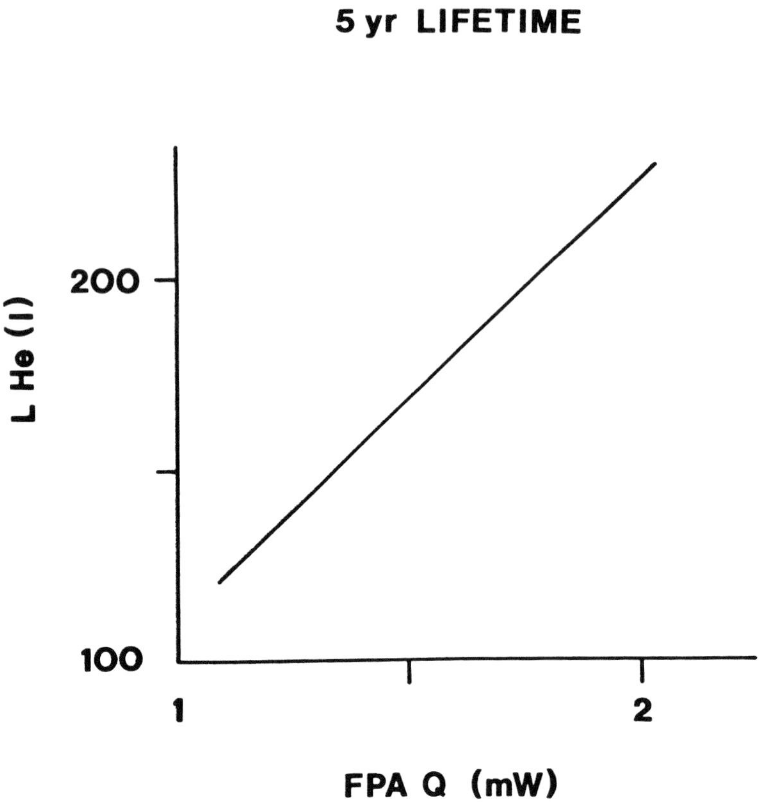

Fig. A2. Amount of liquid helium versus focal plane heat load for a 5-year cryogen lifetime. This dewar size calculation is based on the Ball thermal model which has been validated by test data from the IRAS, and COBE flight dewar systems.

plugs are being investigated. The device will have a small heater for periodic defrosting of the orifice. After the gas has been expanded, it enters an isothermalizing region of sintered metal foam which contains any liquid helium that forms.

The pre-cooling of the J-T valve and focal plane assembly to below the inversion temperature of the helium is accomplished by the use of a bypass line controlled by a room temperature valve. The J-T heat exchanger is a critical item – it pre-cools the high pressure gas to a temperature close to that of the cold gas after returning from the orifice. The gas cleaners remove condensibles from the circulating gas and provide a high surface area at the pre-cooling stages; they also double as heat exchangers. Gas purity will limit the continuous operating period of the J-T system, and it may be necessary to warm up the valve to defrost it. This is not a failure mode as performance will be recovered. The compressors provide a high pressure supply of approximately 12 bar and a low pressure return of 1 bar. There will be a return pressure drop of approximately 0.5 bar on the low pressure side leading to

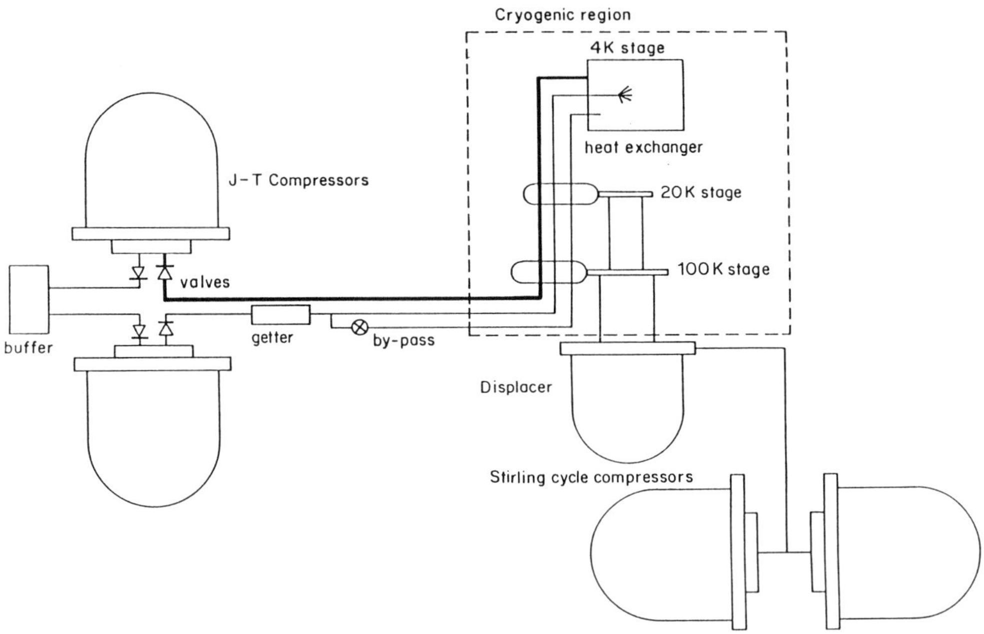

Fig. A3. Conceptual layout of the J-T cooling system which is a cooler option being developed by RAL. The major components of the system are described in the text.

an expansion temperature of 4.6K, since the gas pressure on the exhaust side of the expansion valve determines the base temperature of the cooler. The compressors are very similar to those used on the Stirling cycle cooler with the addition of long life reed valves to produce "dc" pressure. We have tested a compressor with reed valves in the laboratory and find it gives the performance that we require.

ORBITS FOR RADIATIVELY COOLED SPACE TELESCOPES

J.B. FARROW

British Aerospace Space Systems, Earth Observation and Science Division, Bristol

Abstract. This paper first outlines the assumed mission requirements for a radiatively cooled space telescope such as EDISON. A summary of relevant characteristics (payload, operating orbit, launcher, lifetime, etc.) for current and proposed cooled telescope missions is then given. This summary includes cryogenic and radiatively cooled missions since in both cases the reduction of heat input to the telescope aperture is a dominant factor in the orbit choice. These missions span the entire range of possibilities from low earth circular, through higher elliptical and circular orbits out to deep space locations such as the Sun-Earth (S-E) libration points and the lunar surface.

A full listing of the factors affecting mission selection is then given. The most important points are illustrated by reference to the orbits chosen for ISO, FIRST and SIRTF and those recommended in recent studies of EDISON. Launcher capabilities for direct insertion and the onboard propellant for large velocity changes associated with orbit raising are major constraints in achieving the large payload mass to high orbit which EDISON mission requires. Although it is fairly demanding in launch/boost energy, an orbit about the L2 S-E libration point offers important advantages for a radiatively cooled infrared telescope. Further studies of this orbit and the associated aspects of service module and payload design for the L2 location of EDISON are recommended.

1. Requirements

Full details of the requirements for radiatively-cooled infrared telescope missions appear in other papers presented at this workshop. The optimum aperture still has to be decided from considerations of scientific performance at the estimated mirror operating temperature and of technical issues related to payload mass and size, pointing accuracy, etc. As a basis for this assessment of orbit choices, the telescope aperture range (which strongly influences the dry spacecraft mass) was assumed to extend from the shared Ariane-4 'POIROT-Class' (Reference B9) at 1.5m diameter up to a dedicated Ariane-4 or Atlas 2 class of mission at 2.5m diameter. Typical configurations at the limits of this size range are shown in Figure 1.

From discussion at the workshop it is clear that there is strong support for apertures greater than 2.5m, up to the limits imposed by launcher shroud diameters on a monolithic mirror. The more modest 2.5m limit assumed that the corresponding 'dry' spacecraft mass and the associated fuel would lie on the limit of the capability of 'affordable' launch vehicles for the likely choice of orbit. Clearly these factors – spacecraft total mass, launcher availability and location – are interdependent as described in the remainder of this paper.

The assumed requirements are listed in Table I. Note that the spatial resolution of a 2.5m telescope at 2 μm is about 0.2 arc sec and to obtain diffration-limited performance at such short wavelengths would require sub arc second level pointing. Very high pointing attitude control was identified as a key design driver at the workshop. (BAe has recently completed an ESA-funded study of this subject

Space Science Reviews **61:** 187–210, 1992.

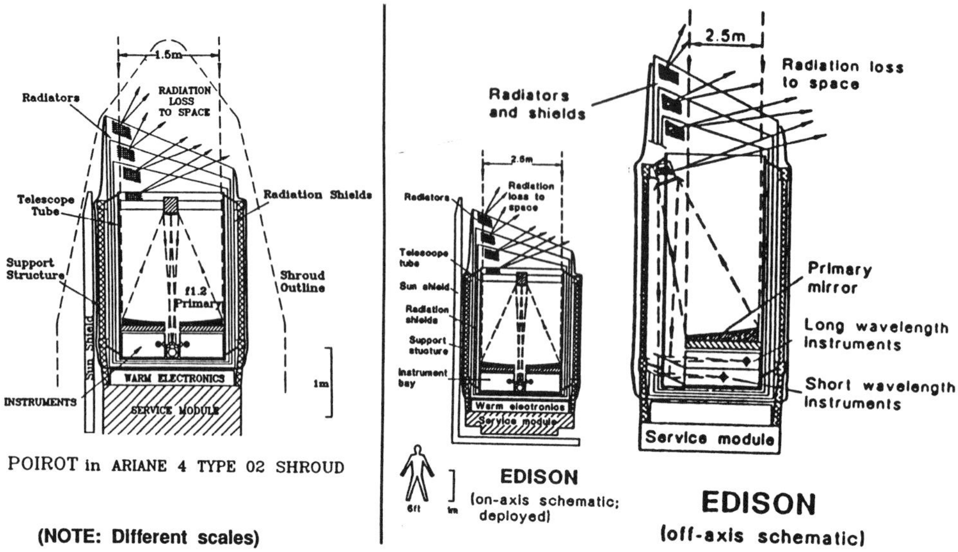

Fig. 1. Configuration examples – POIROT and EDISON.

which addressed factors such as component performance limitations, fine error sensing, boresight alignment, etc., taking FIRST as a model mission).

It is understood that the raw scientific data rate is likely to be several hundreds of kilobits per second (or even a few Mbps) for uncompressed data from large area detector arrays with fast frame rates. For a mission launch date after the year 2000 it is considered reasonable to assume a modest level of onboard data compression to restrict the processed data rate to a few hundred kilobits per second.

An 'observatory' mode of operation, allowing near real-time interactive control for a high percentage of the time , is clearly a very desirable feature of a mission such as EDISON. Data retrieval considerations obviously have an important influence on the orbit choice though these aspects will be only briefly addressed in this paper.

It will be seen that mass rather than size presently establishes the maximum usable telescope aperture diameter. The limit of about 2.5m assumed here is based on conventional approaches to mirror fabrication and support. Further advances in the development of lightweight, possibly active structures has been identified as another technical goal which would clearly allow further increases in aperture size for a given limit in launch mass.

2. Relevant Missions

It is not the intention of this paper to provide a detailed review of cryogenically and radiatively-cooled infrared telescopes which have been developed or proposed elsewhere. However, since all such missions have to take into account the main factors affecting orbit selection, it is interesting to note what choices have been

TABLE I
Main requirements for radiatively-cooled IR telescopes

FEATURE	TYPICAL VALUES
• Low (uniform) temperature	30-60K for optics (< 10K for detectors)
• Large aperture	1.5-2.5m (or greater)
• Wide spectral range	2-50 μm
• Long lifetime	> 5yrs
• High observing efficiency	> 70%
• Background radiation level	Low
• Pointing accuracy/stability	Arc second level
• Data rate	Several 100s of kbps
• Launch date	> 2000

made by other project planners. Figure 2 and Table II provide relevant information on several cryogenically cooled missions:-

- IRAS – Infrared Astronomy Satellite (launched in 1983)
- ISO – ESA's Infrared Space Observatory (to be launched ~ 1994)
- STEP – Satellite Test of the Equivalence Principle (a proposed medium-class M2 ESA mission)
- COBE – Cosmic Background Explorer (launched in 1989)
- SIRTF – Space Infrared Telescope Facility (under development for launch after the late 1990s)

Full details of these missions can be found in References A1-A6.

Figure 3 and Table III provide similar information on radiatively cooled missions (described fully in References B1-B9) including:-

- FIRST – Far IR and Submillimetre Telescope (an ESA cornerstone mission due for launch after 2002).
- POIROT – A Passively Cooled Orbiting Infrared Telescope (a mission proposed, but rejected, for M2).
- SWAS – The Submillimetre Wave Astronomy Satellite (a NASA mission proposed for a mid-90's launch).
- SMMM – Submillimetre Mission (a US/French mission proposal for a late 90's launch).
- LDR – Large Deployable Reflector (a proposed US mission for launch after 2000).

TABLE II

Main characteristics of cryogenically cooled missions

MISSION	PAYLOAD CHARACTERISTICS			OPERATING ORBIT			LAUNCH / BOOST	LIFE Yrs.	LAUNCH MASS, Kg
	Dcm	λ µm	He, l	Thrs.	h Km	i deg			
<u>EUROPEAN</u>									
IRAS (1983)	60	8 - 120	500	1.7	900	99 (S.S.) †	Delta 3910	~1	800
ISO (1993)	60	3 - 200	2,300	24 *	1000 x 70,000	5 - 20 (TBD)	Dedicated Ar 4+ Hydrazine RCS	> 1.5	2,400
STEP (2000)	(Drag - free probe)		200	1.6	550 - 690	98 (S.S.)	4 - stage Taurus	> 0.5	820
<u>U.S.</u>									
COBE (1989)		1 - 10,000	600	1.7	900	99 (S.S.)	Delta	1	
SIRTF (> 2000)	95	2 - 700	4,000	96 **.	100,000	28.5	Titan 4 / Centaur	>5	4,500

† S.S. = Sun Synchronous.
* See later sheet concerning selection of ISO's orbit.
** See later concerning selection of SIRTF's orbit.

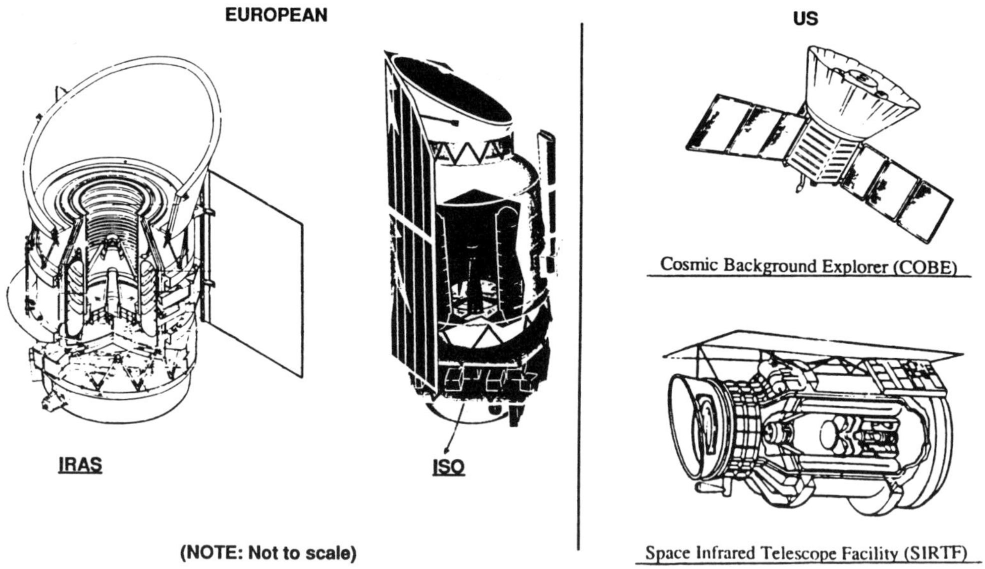

Fig. 2. Illustrations of cryogenically cooled telescope missions

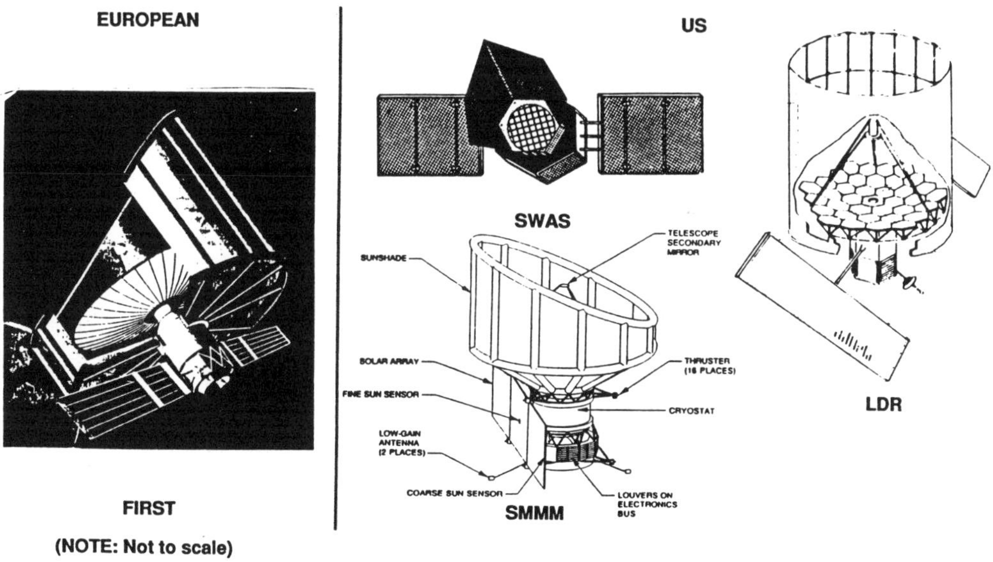

Fig. 3. Illustrations of radiatively cooled telescope missions.

TABLE III

Main characteristics of radiatively cooled missions

MISSION	PAYLOAD CHARACTERISTICS			OPERATING ORBIT			LAUNCH / BOOST	LIFE Yrs.	LAUNCH MASS, Kg
	Dm	λ μm	Tk	Thrs.	h Km	i deg			
EUROPEAN									
FIRST (2002)	4 (-8)	100 - 1000	150	24	1000 x 70,000	7°	Shared Ar 5 into GTO + Hydrazine RCS	6	3,400
POIROT (2000)	1.5	2 - 40	40 - 50	24	1000 x 70,000	7°	Shared Ar 4 + boost from GTO	>5	1,950
U.S.									
SWAS (1994)	0.55 x 0.7	550 - 620		1.6	550 - 600	0°	Scout	>2	200
SMMM (late 90's)	3.7	100 - 700		24	1000 x 70,000		Atlas (or Delta with ø = 2.5 m)	2	
LDR (> 2000)	10 - 20	50 - 1000	150 - 200	96	100,000	28°	TBD	8 - 10	V. Large
OTHERS									
MOSES (Sweden, etc.)	<1	~500		1.6	TBD	Near Polar S.S.	Pegasus	TBD	Low
RELICT - 2 (USSR, 1992)	μwave receivers			Orbit about L2 point					

3. Orbit Options

Figure 4 shows possible orbits about the Earth and beyond in the form of an options tree diagram. The approximate location of the various missions summarised in the previous section is indicated in the diagram. Two main points emerge. First, almost every possible class of orbit has been considered for cooled infrared telescope missions. Popular choices have been:-

- Near-polar sun synchronous e.g. IRAS, COBE, MOSES low earth orbit (LEO)
- 24 hr elliptical orbit e.g. ISO, FIRST, POIROT, SMM
- High circular orbit (96 hr) e.g. Baseline SIRTF, LDR
- Libration points of the Sun-Earth System (L1, L2) e.g. RELICT-2, SIRTF alternative.

Second, other mission planners have found difficulty in deciding upon the optimum orbit as illustrated by the following examples:-

- ISO was assumed to occupy a 12 hr elliptical orbit (1000km perigee, 40,000km apogee) until well into its development phase when the recommendation to increase its apogee height to 10,000km (24 hr) was adopted.
- SIRTF has been studied for very many years but only recently has the decision been made to change from LEO (900km circular) to a high circular orbit at 100,000km (96 hr period). Some consideration has also been given to the possibility of placing SIRTF at the L2 S-E libration point.
- POIROT was proposed as a shared Ariane-4 passenger which would then have been boosted from geostationary transfer orbit (GTO) to a 24 hr period ISO-like orbit. Subsequent more ambitious versions of the EDISON-class have considered elliptical orbits with higher perigee and apogee as well as a high circular orbit (100,000km).

Some of the reasons surrounding the selection of these and other possible orbit choices will be outlined in the remainder of this paper.

4. Key Factors in Orbit Selection

4.1. Introduction

Key factors influencing the choice of orbit for any astronomical mission are listed in Table IV against four main categories:-

- Scientific Objectives/Performance
- Launch and Operations
- Data Retrieval
- Spacecraft Design

For radiatively-cooled missions the attainment of a low, stable telescope temperature is the primary objective. Consequently every effort must be made to minimise and stabilise the heat input to the telescope. This is clearly best achieved by placing the spacecraft in a 'high' orbit where the mean inputs due to earth albedo and emitted infrared radiation, and variations in these inputs around the orbit, are

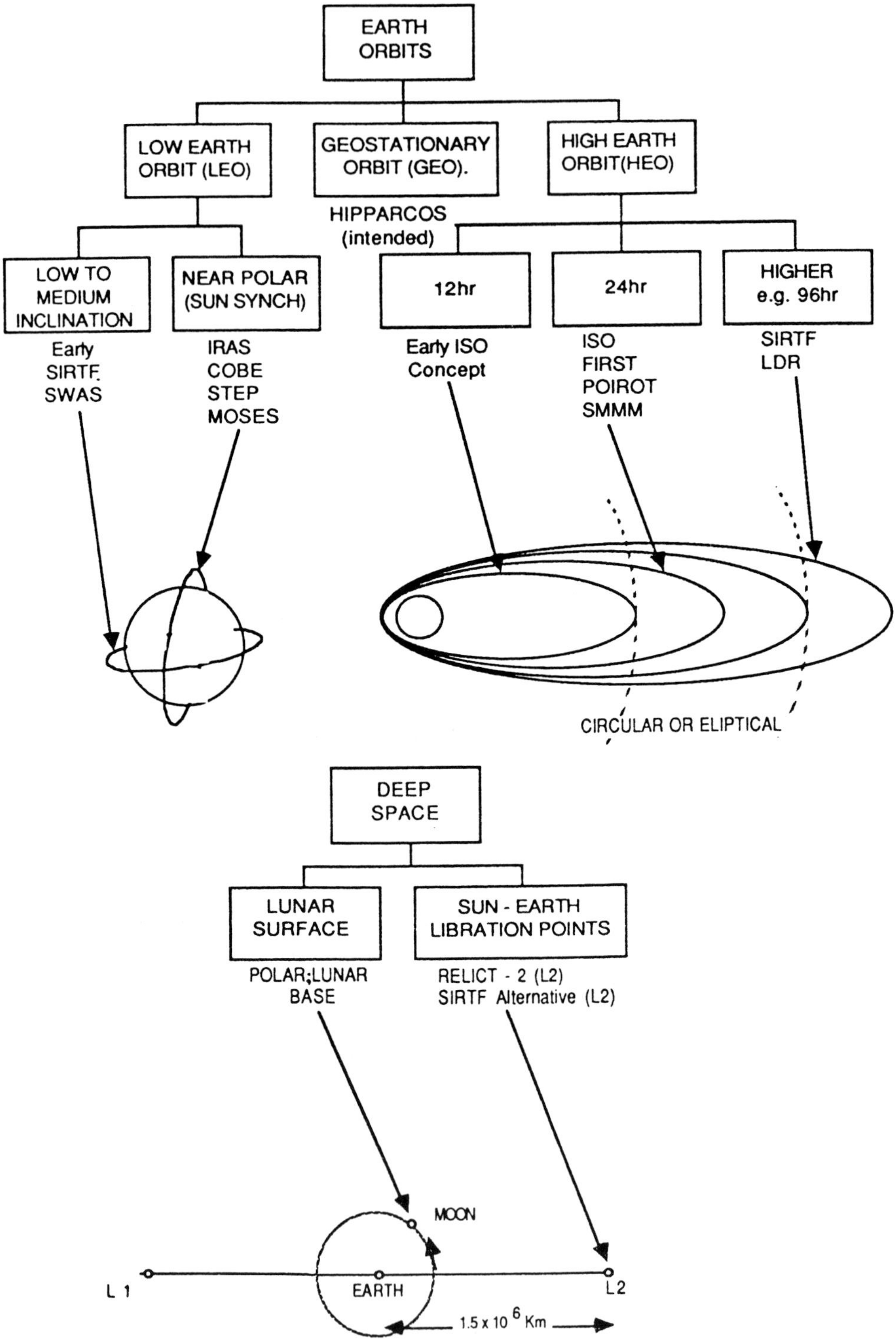

Fig. 4. Orbit options diagram.

minimised. Operating features of the telescope (such as its observing efficiency, the freedom to view faint targets for long uninterrupted periods, avoidance of earth occultation, etc.) are also greatly improved by the adoption of high rather than low earth orbits.

The main attractions of a near-polar LEO, which led to its selection for IRAS, COBE, etc., were:-

- Large spacecraft mass into orbit;
- Simple spacecraft design (in a dawn-dusk, sun synchronous orbit in which the sun always lies near normal to the orbit plane);
- Good sky survey capability using a viewing direction aligned with an earth radius vector.

It is understood that a radiatively-cooled telescope in LEO would stabilise at an equilibrium temperature above 100K (c.f. IRAS and COBE after all of their cryogen had evaporated). Whilst such a temperature might be acceptable for a modest-aperture, 'precursor' mission it is assumed to be unacceptable for EDISON. Thus low earth orbit operation is not considered further in this paper, but the merits of the proposed Japanese small mission described by Hayashi at the workshop are recognised.

Background radiation severely impedes scientific measurements during passage through the Earth's trapped electron and proton belts and the accumulated radiation dose has important impacts on semiconductor components (e.g. IR detectors, spacecraft electronics, solar cells). For these reasons the radiation belts are best avoided by staying below them ($h < 1000$ km), remaining above them ($h > 40{,}000$ km) or at least by minimising the number of passages through them (i.e. by selecting a long period elliptical orbit). The geostationary orbit (GEO 36,000 km circular) is unacceptable on these grounds in spite of its other attractions of continuous contact with a single ground station and well developed spacecraft bus equipment.

Having eliminated LEO and GEO from the initial set of options it is now appropriate to look in more detail at the influence of the selection factors of Table IV on various high earth and deep space orbits.

4.2. Thermal Input

The temperature at which a radiatively-cooled telescope will stabilise depends on:-

- Internal factors (power dissipation in focal plane; conductance and radiation effects, etc.);
- External factors (incident flux due to the three main components illustrated in Figure 5).

The contribution of direct solar flux, earth albedo and earth IR emission will depend on: –

- The orientation of the spacecraft with respect to the sun and earth;
- The height of the spacecraft above the earth which directly governs the solid angle as also shown in Figure 5;

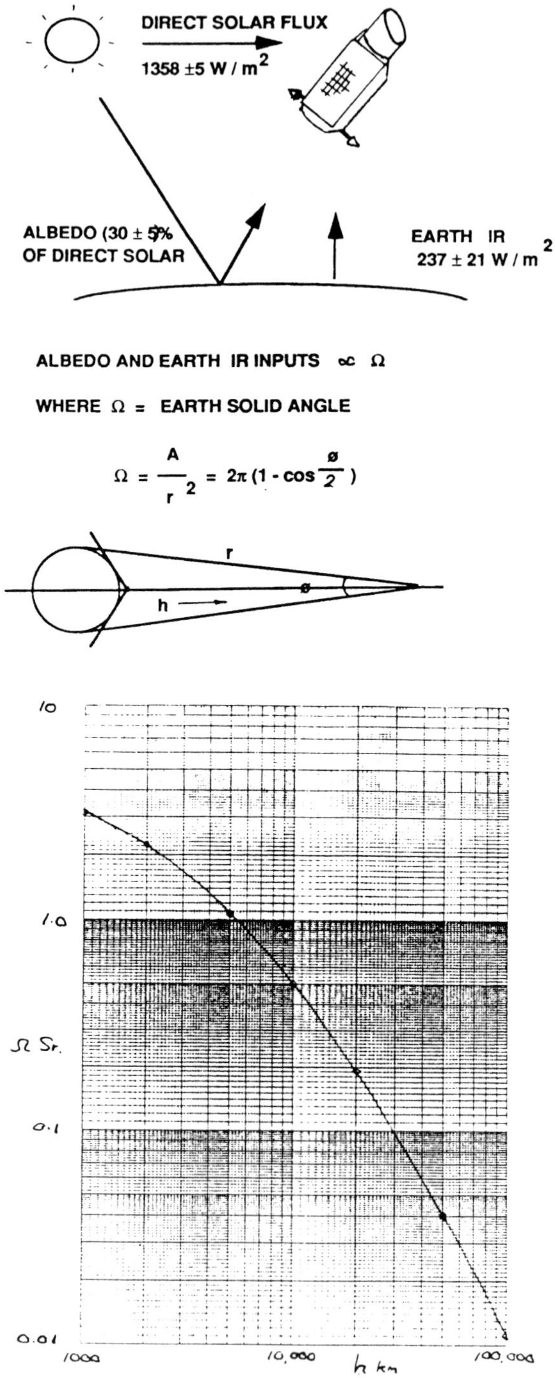

Fig. 5. Thermal inputs.

TABLE IV
Key factors in orbit selection

Factor	Preference
• Scientific Objectives / Performance	
TABLE IV Key factors in orbit selection	
– Thermal Input (Magnitude, Variation)	Generally Favour HEO (or beyond)
– Viewing Efficiency (Earth Occultation, etc.)	
– Celestial Sphere Coverage	
– Uninterrupted Observing Time	
– Operating Mode (e.g. 'Observatory-Like')	
– Background Radiation Noise	
– Contamination	
Key factors in orbit selection	
• Launch and Operations	
– Launcher Availability, Capability and Cost	LEO
– Scope for Sharing Launch	HEO (via GTO)
• Data Retrieval	
– Ground Station Coverage / Cost	GEO
– Range / Angle for Communications	LEO/GEO
– Onboard DH Complexity / Reliability	GEO
• Spacecraft Design	
Key factors in orbit selection	
– Thermal Control	Depends on Specific Parameters
– Eclipse Duration / Frequency	
– Orbit Raising / Maintenance	
– Solar Array Sizing	
– Attitude Control System, etc.	

– The thermal properties of the spacecraft surfaces;

– The shielding effects of the structure, limiting the incidence of sun and earth radiation into the telescope aperture.

The same factors also dictate the boil-off rate for cryogenically cooled missions, of course, and so the orbit considerations are broadly similar for the two cases. Hence the orbit trade-offs and optimisations conducted for ISO, FIRST, SIRTF, etc. are directly applicable to EDISON if one replaces cryogen boil-off rate by mirror temperature as the key design criterion.

The rapid decrease in earth heat input with altitude is clear from Figure 5. For example:-

- Raising SIRTF's orbit from circular LEO at 900 km to a high circular orbit at 100,000 km gives a constant reduction in solid angle of about 300; the dominant effects of earth heat inputs in LEO are then reduced to a level of only 'minor importance' with a greatly extended cryogen life.
- Raising ISO's apogee height from 40,000 to 70,000 km reduces the number of perigee passages, where maximum earth heat input is received, from two to one per day giving a useful (approximately 10%) increase in cryogen lifetime.

Predictions of the actual mirror temperature which will be achieved radiatively for a particular orbit and telescope geometry demand the use of thermal analysis models of the type described by Hawarden *et al.* in these proceedings. The paper by Hawarden *et al.* took the example of a high circular orbit at 100,000 km. It is recommended that other cases be analysed including for example:-

- Elliptical orbits with various perigee and apogee heights, since those offer significant savings in launch/boost energy;
- Libration point orbits since these have other advantages as described later.

For elliptical orbits it is necessary to assess the dynamic thermal behaviour around the orbit, following the main earth input which occurs during the relatively short perigee passage.

4.3. Elliptical Orbits

Figure 6 illustrates a family of elliptical orbits chosen, for convenience of ground station operation, to be multiples of an earth rotation period. A shared launch on Ariane 4 (or 5), injects the spacecraft into a geostationary transfer orbit (GTO) which has perigee and apogee heights of 200 km and 36,000 km i.e. it lies just within the ellipse labelled '12hr'. Tick marks are drawn every 2 hrs around the orbits. Observatory operations are suspended during the perigee passage partly because of the high spacecraft angular rate with respect to the ground station but, more importantly, because of the high radiation environment. Energetic protons dominate at the lower altitude range, down to about 1000 km; the effects of trapped electrons are significant up to 40,000 km or even higher. However, as can be seen in Figure 6, a spacecraft in a 24 hr ISO-like orbit spends 70% of the time above this height.

The original 12hr ISO orbit (1000 km x 40,000 km) would have been achievable by a shared launch on Ariane 4 followed by only a modest amount of orbit raising using the spacecraft propulsion system. The change to a 24hr period orbit offers significant advantages through:-

- A reduction in cryogen boil-off rate (see Section 4.2);
- A reduction in the upper baffle average temperature and hence a lower level and an improved stability of thermal 'straylight';
- Reduction in the effects of background radiation flux (see Figure 6);
- An increase in observing efficiency;
- Some reduction in spacecraft design and development risk.

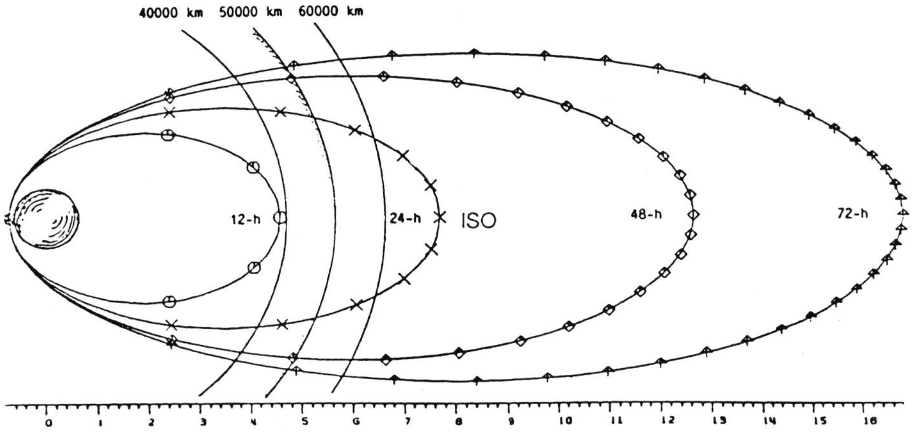

12-, 24-, 48- and 72-hour orbits with tick marks every two hours and circles of constant altitude for 40000, 50000 and 60000 km.

Fig. 6. HEO orbit illustrations.

The main drawback was the need for a dedicated rather than a shared Ariane 4 launch, adding significantly to the cost of the launch and hence the total programme cost.

Referring to Figure 6 again, it is clear that increasing the apogee height, and hence the period, still more would give further improvements in each of the areas listed above but at the expense of even greater launch/boost energy.

For astronomical missions in elliptical orbit the perigee height is normally chosen to lie in the range 500-1000 km. This reduces drag effects during the perigee passage to a low level whilst optimising the payload mass into the final operating orbit. However, for radiatively-cooled telescopes a significantly higher perigee may be required to:-

- Reduce the significant level of earth heat input in this part of the orbit;
- Reduce the mass and size of the earth shield;
- Avoid the worst effects of the radiation belts.

For these reasons the EDISON working group have briefly considered a 20,000 km perigee and a high apogee e.g. approximately 100,000 km (48hr) or 140,000 km (72hr). During the perigee passage the earth solid angle (Figure 5), and hence the heat from this source, would be 16 times less than for the 1000 km perigee assumed in the POIROT proposal. Hence the telescope would stabilise at a significantly lower, but still to be determined, temperature. Furthermore in a 72hr orbit the number of perigee crossings would be three times less than in the 24hr orbit with corresponding reductions in the cumulative thermal and radiation effects.

5. Launch Vehicles

A separate paper in this volume addresses the question of launch vehicles (and spacecraft design) in some detail. To illustrate the problems of placing spacecraft of large mass into high orbit, Table V provides a summary of the performance capabilities of Ariane 4 and 5, Delta 2 and Atlas launchers. The paper by Turner and Curtis includes details of even more powerful launchers such as Titan and Energia. From Table V it is clear that the mass that can be placed directly into orbit falls off as one moves from LEO, through GTO and on to HEO. The actual performance is a function of launcher type (e.g. number of strap-on boosters), desired altitude and inclination. Figure 7, for the particular case of an Ariane 44L launch at an inclination of 7°, is typical of the performance variation as a function of perigee and apogee altitude.

6. Orbit Raising

6.1. Velocity Change for Orbit Raising from GTO

Figure 7 shows that a dedicated Ariane 44L can place a spacecraft mass of 4200kg into a geostationary transfer orbit. In a shared launch mode as proposed for POIROT, rather less than half of this total (1950kg after allowing for the mass of the separation structure) can be allocated to each of two similar-sized satellites. Figure 8 shows the velocity change that would then be required to raise perigee and apogee heights starting from GTO. Values appropriate to 24, 48 and 96hr elliptical orbits with 1000 km perigee and for geostationary orbit are highlighted.

To calculate the effect of imparting such velocity changes on the useful payload mass we use the equation

$$M_{PL} = [M_i \exp(-\Delta V/gI_s)] - M_{sm} \qquad (1)$$

where
M_{PL} = payload module mass (i.e. telescope, focal plane assembly etc.),
M_i = total mass placed in initial orbit (e.g. 1950 or 4200kg for shared or dedicated Ariane 44L launch into GTO at 7°),
ΔV = required velocity change from initial to final orbit,
I_s = specific impulse of propulsion system used for orbit raising,
M_{sm} = service module dry mass (plus allocation of fuel for attitude control).

As a rough approximation we may take a value of 730kg for M_{sm}, as suggested by ESA for the 'common astronomy bus' concept design. Figure 9, again based on ESA guidelines, shows how the payload mass which could be placed into GTO (1950-730 = 1220kg) then falls off with further increases in perigee and apogee height. These curves are based on a specific impulse of 220 sec appropriate to a hydrazine reaction control subsystem.

It is worth noting that for large ΔV changes the useful payload could be much improved by using a high specific impulse system such as an ion motor. Electric

TABLE V

Performance summary for expendable launch vehicles

VEHICLE	APPROX. PERFORMANCE INTO :- LEO †	GTO (200 x 36,000Km)	24hr. HEO (200 x 70,000 Km)	FAIRING DIA. m	APPROX. LAUNCH COST. MAU (90)
EUROPEAN					
ARIANE 4					
- SINGLE PAYLOAD	4,850 - 9,550	4,200 (44L)	3,600	3.65 max.	100 - 120
- SHARED PAYLOAD	*	$\frac{3,900}{2}$ = 1950	*		60 - 95
ARIANE 5 (> 1996)					
- SINGLE PAYLOAD	10,000 - 18,000	6,800	TBD	4.57 max.	TBD
- SHARED PAYLOAD	*	5,900			
U.S.					
DELTA 2	3,800 - 5,000	1,450 - 1, 810 †	1,250 - 1,550	2.5 / 2.8	43 - 48
ATLAS 1	6,000 - 8,600 FOR i = 28°	2,300	1,900	3.4 - 4.2	62 - 67
ATLAS 2		2,700	2,300		67 - 76
ATLAS 2A		2,800	2,400		76 - 86
ATLAS 2AS		3,450	3,000		105 - 114

† FUNCTION OF LAUNCHER TYPE, ALTITUDE AND INCLINATION   * NEGOTIABLE DEPENDING ON PARTNER

propulsion offers a 10-fold improvement in I_s compared to conventional mono or bipropellant systems but there are some important drawbacks for an EDISON-like mission:-

– A significant increase in the required solar array size to provide the power necessary for the ion motor;

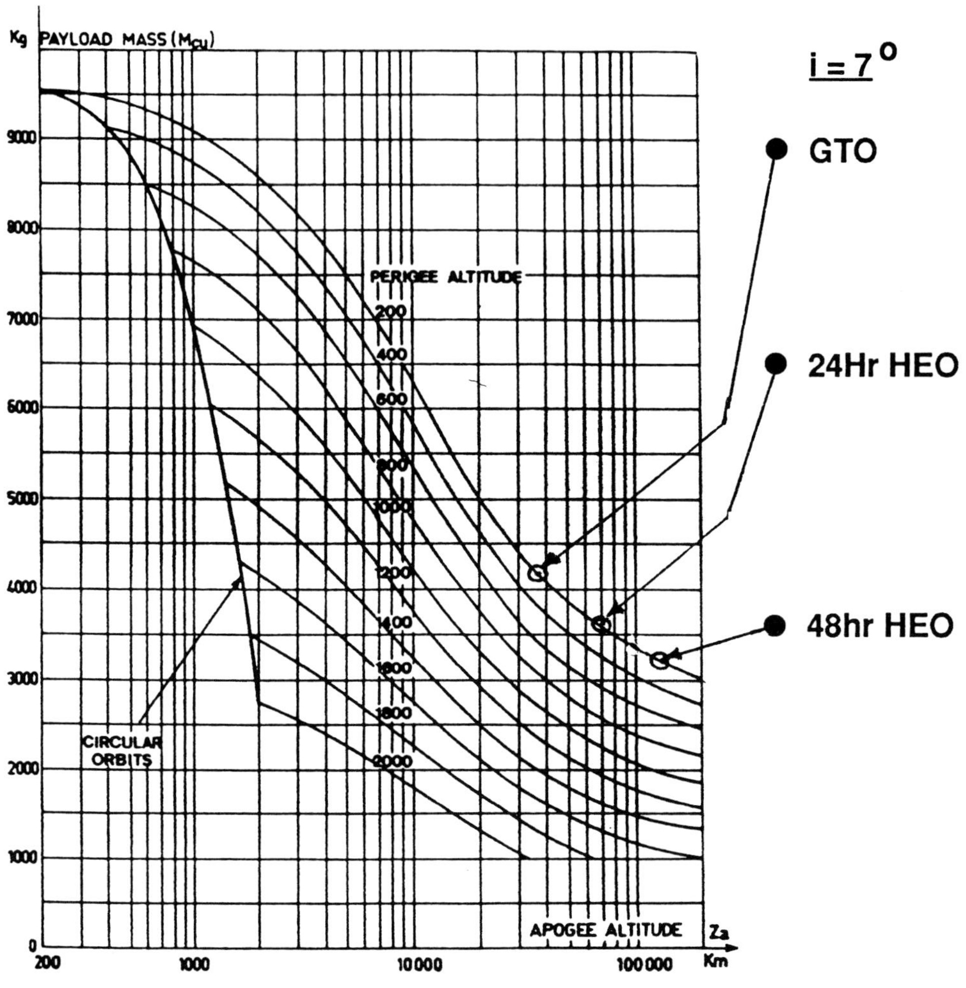

Fig. 7. Ariane 44L performance.

– A very long orbit transfer time of many months for the much lower thrust level of electric propulsion systems; this may not be such a problem if it takes several months for the telescope to reach its final operating temperature.

From Figure 9 it is seen that with a hydrazine reaction control system, a payload mass of 900kg could be achieved in a 24hr ISO-like orbit after a shared Ariane 4 launch into GTO. This was the case for the proposed POIROT mission based upon a 1.5m telescope aperture. With a dedicated Ariane 4 launch, the above equation shows that a payload mass of 2775kg corresponding to a 2.5m class telescope could be raised to this 24hr orbit. Obviously a similar on-orbit mass can be achieved by

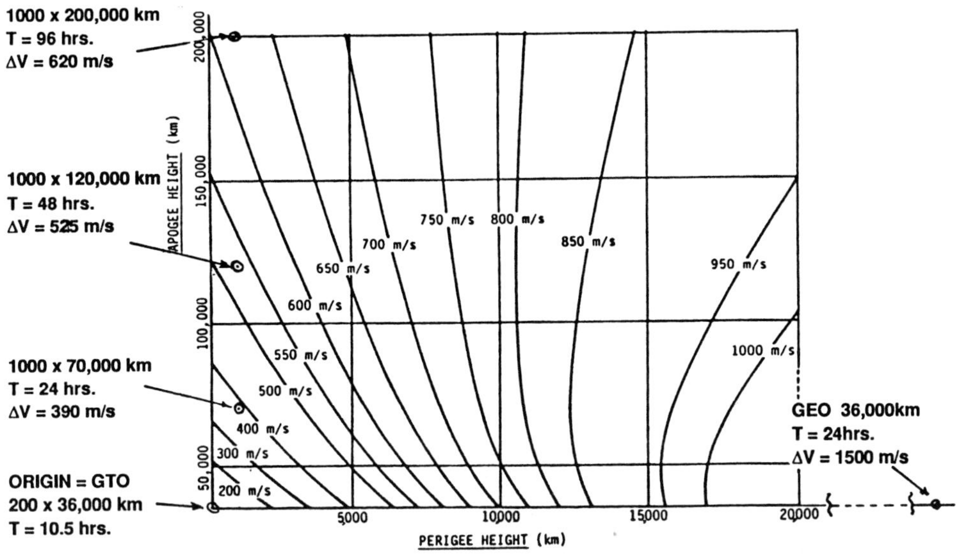

Fig. 8. Velocity change for orbit raising from GTO.

using the dedicated launch to insert the spacecraft directly into an orbit with the desired apogee height (as shown in Figure 7).

6.2. High Circular Orbits

As discussed earlier, high circular orbits are much favoured for the mission in question in scientific performance terms, because of the large reduction in earth heating and background radiation. The drawback is that starting from GTO, or other initial orbits in the case of a dedicated launch, the ΔV to achieve a high circular orbit is greatly increased. This effect is indicated at the top of Figure 9 where the resulting payload mass is seen to be impractically low for a shared Ariane 4 launch. Table VI lists the ΔV required for various high circular orbits when starting from GTO with:-

ΔV_1 – apogee raising velocity change from GTO to HEO;

ΔV_2 = velocity change to circularise the high orbit;

$$\Delta V = \Delta V_1 + \Delta V_2$$

The total ΔV is seen to change only very slowly above 60,000km. In the last column of Table VI the fuel mass, expressed as a fraction of launch mass, assumes a specific impulse of 290 sec. The total Ariane 4 GTO mass capability reduces to a payload mass of only (0.55 x 4200)–730 = 1580kg in these high circular orbits. To place a 2.5m class telescope there (assumed to correspond to M_{PL} = 2800kg)

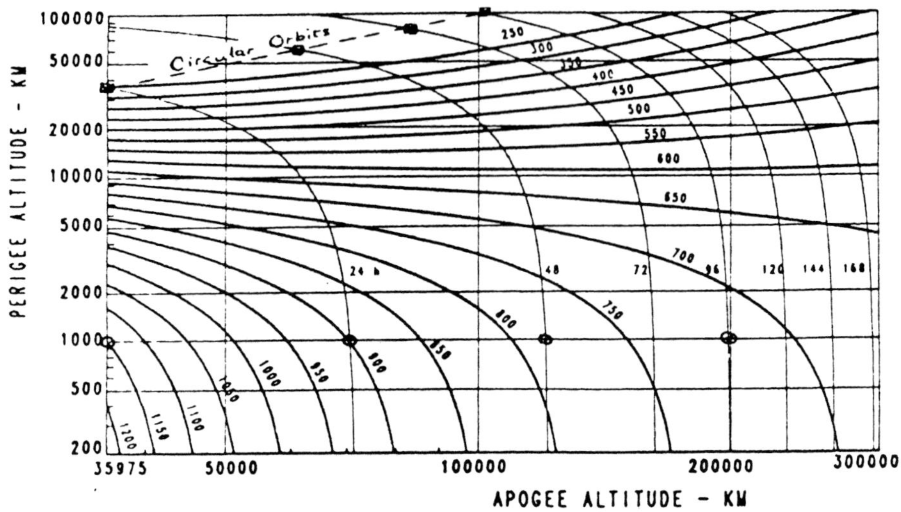

ORBIT Km	ΔV FROM GTO, m/s	Mp/l Kg
200 x 36,000 ; GTO	0	1220
1000 x 36,000 ; 12hr	75	1150
1000 x 70,000 ; 24hr	390	900
1000 x 120,000 ; 48hr	525	800
1000 x 200,000 ; 96hr	620	730

Fig. 9. ESA nomograms for shared launch plus boost.

TABLE VI
ΔVs for high circular orbit

Final Circular Orbit Height, Km.	Period, hr	ΔV_1 m/s	ΔV_2 m/s	Total ΔV m/s	Fuel Mass as Fraction of S/C Launch Mass*
36,000 (GEO)	24	–	1480	1480	0.41
61,000	48	290	1400	1690	}
100,000 (SIRTF)	96	424	1295	1719	} 0.45
150,000	170	536	1158	1694	}

* Assuming I_s = 290 sec.

would require a more powerful launcher (e.g. Ariane V or Titan) at a significantly higher cost.

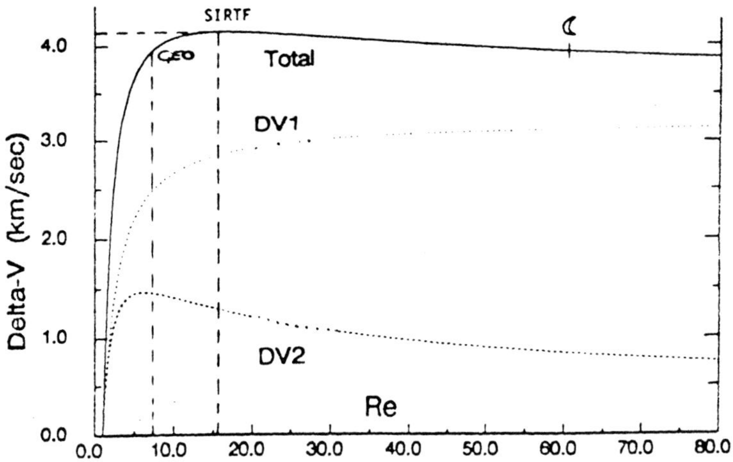

Fig. 10. ΔV for high circular orbit (Ref C3).

Figure 10 shows the total ΔV required to achieve a high circular orbit, expressed in earth radius units, when starting from a low circular orbit at 185 km. Again ΔV_1 and ΔV_2 refer to the velocity changes for apogee raising to the desired altitude and circularisation at this altitude respectively.

Points corresponding to GEO to SIRTF's 100,000km circular orbit and to the moon's orbit are indicated in Figure 10 and these are seen to have similar ΔV requirements.

6.3. Sun-Earth Libration Point Orbits

At high Re the value of ΔV approaches 3.2 km/s, the required escape velocity from a low circular orbit. It then becomes possible to enter a cruise trajectory towards a deep space orbit. Figure 11 shows an example trajectory to the L2 libration point of the Sun-Earth (S-E) system i.e. one of the points at which a spacecraft would be in equilibrium under the centripetal and gravitational forces acting upon it. The S-E L1 point, also shown in Figure 11, has already been adopted for NASA's ISEE-3 mission and chosen for ESA's SOHO mission. L1 at a distance of 1.5 x 10^6 km from Earth in the direction of the Sun is an ideal location for those missions as it lies upstream of the Earth's magnetosphere with constant solar illumination and at a fixed range. The libration point itself is only quasi-stable and to avoid large ΔV changes for orbit insertion and maintenance a wide amplitude 'halo' orbit about the point itself is favoured.

An orbit about L2, a similar distance from Earth as L1 but on the anti- sunward side, would be preferred for stellar observations since it allows an unhindered view of over half of the celestial sphere at all times. Perturbing sources of stray radiation (Sun, Earth and Moon) all lie on the same side of the spacecraft. Lower amplitude (Lissajous) orbits could be considered to reduce the range of angles to the earth

TABLE VII

Summary of advantages and drawbacks of various locations

CHARACTERISTIC	LEO	GEO	HEO	HIGH CIRCULAR	L2	LUNAR (POLAR)
ASSUMED ORBIT HEIGHT Km	900 (S.S.)	36,000	1000 x 70,000	100,000	1.5×10^6	-
TELESCOPE TEMPERATURE, K	~100	40 - 50	40 - 50	<35	<35	40
THERMAL STABILITY	POOR	GOOD	FAIR	GOOD	GOOD	GOOD
OPERATING EFFICIENCY, %	45	90	>70	90	90	90
INST. SKY COVERAGE, % *	< 10	~30	~30	~30	>60	<50
UNINTERRUPTED VIEWING TIME	SHORT	LONG	LONG	V LONG	V LONG	V LONG
SCIENTIFIC OPERATIONS	COMPLEX	SIMPLE	FAIRLY SIMPLE		SIMPLE	COMPLEX
BACKGROUND RADIATION	POOR (S.A.A.)	POOR	FAIR	GOOD	GOOD	FAIR
LAUNCHER CAPABILITY / COST	GOOD	POOR	FAIR	POOR	POOR	BAD
SHARED LAUNCH PROSPECT	POOR	FAIR	GOOD	FAIR	POOR	-
GROUND STATION COV. /COST, OBDH / COMMS COMPLEXITY	POOR	GOOD	FAIR	FAIR	FAIR	FAIR
S/C DESIGN COMPLEXITY						
- THERMAL CONTROL	FAIR	FAIR	GOOD	GOOD	GOOD	HIGHLY COMPLEX/ COSTLY INFRA - STRUCTURE
- PROPULSION	GOOD	FAIR	FAIR	FAIR	FAIR	
- POWER & SOLAR ARRAY	GOOD (DAWN-DUSK)	FAIR	FAIR	GOOD	GOOD	
- ATTITUDE CONTROL	POOR	FAIR	FAIR	FAIR	GOOD	
COMMON BUS PROSPECT	POOR	FAIR	GOOD (XMM)	FAIR	FAIR / GOOD	
SUITABLE FOR EDISON	NO	NO	PERHAPS	YES	YES	NO

* DEPENDS ON RELATIVE LOCATION OF SUN, EARTH, MOON, S/C.

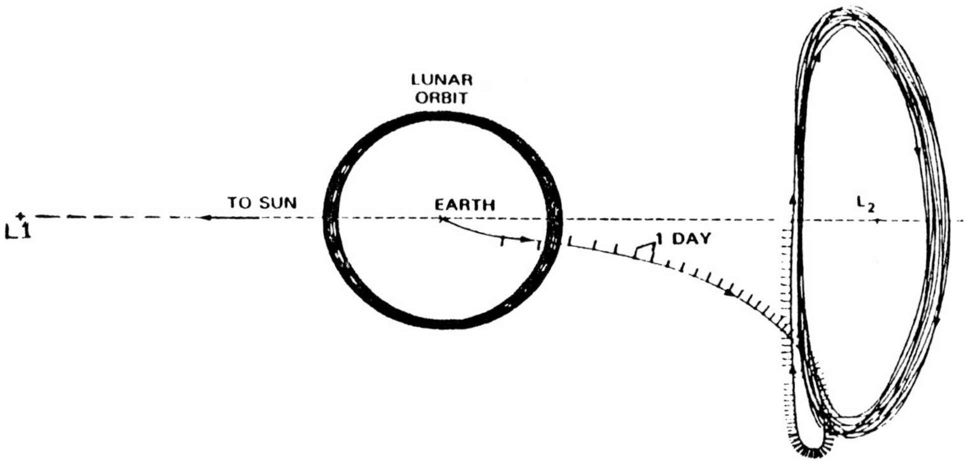

Fig. 11. Halo orbit about L2.

during each orbit period around L2 which lasts several months. An L2 orbit has been selected for the USSR RELICT-2 microwave mission (Ref B6).

For an EDISON type observatory an L2 orbit would be an excellent choice from the point of view of performance since:

- Very low (<35K) stable mirror temperatures should be achievable;
- High operating efficiency would be guaranteed (assuming adequate ground station coverage by 2 or 3 stations, or the use of high capacity recorders if the number of stations is limited);
- Instantaneous sky coverage would exceed 60% of the celestial sphere, the entire sky being accessible in 6 months as the Sun-Earth line rotates.

In these respects an L2 orbit is even better than the high circular orbit proposed for SIRTF. The launch/boost energies required to achieve libration point orbits are high compared to elliptical earth orbits but are no worse than for the high circular orbits. Indeed significant savings can be achieved as these examples of alternative routes to L2 show:-

- An earlier launch mode considered for SOHO (to L1) used an Atlas vehicle for injection into circular LEO and a Centaur upper stage to provide the 3.2 km/s boost into an escape trajectory – equivalent to DV1 at high Re in Figure 10. Subsequent velocity changes, provided by the spacecraft propulsion system, amounted to only about 250m/s (for mid-course manoeuvres, halo orbit insertion and maintenance) significantly lower than the circularisation DV2 in Figure 10.
- In the current SOHO launch mode an Atlas 2AS vehicle will place the spacecraft in an elliptical earth orbit. Boost into the transfer orbit and all subsequent velocity changes will use the onboard propulsion system. The escape velocity from GTO, as an example, is about 750m/s. Assuming 250m/s for subsequent

manoeuvres leads to a spacecraft fuel capacity equivalent to a total velocity change of about 1 km/s which is significantly lower than the ΔV figures given earlier in Table VI to achieve high circular orbits.

- By suitable targeting of the spacecraft towards the moon, *en route* to L2, lunar swingby can be used to further reduce the required velocity change (or to achieve a lower halo orbit amplitude about L2). As pointed out by Farquhar and Dunham (Ref C3) the gravitational force of the sun itself reduces the total ΔV requirement by effectively providing the necessary circularisation at L2.

The paper by Turner and Curtis provides further details of the SOHO service module design which could be very relevant and cost-effective for an EDISON-like mission in certain subsystem areas e.g. structure, propulsion, data handling, communications and power. EDISON's required pointing range, towards stars rather than the sun, would obviously demand modifications to the attitude control and solar array subsystems and thermal control would be both different and more demanding.

7. Lunar Observatory

The advantages of the moon as an astronomical site have been recognised and well described in many recent publications (see references D1 - D3 for example). To achieve the low mirror temperatures required for EDISON it would be necessary either to place the telescope in a deep shadowed crater close to a lunar pole or to rely upon sophisticated thermal shielding and control methods in a more convenient location. The polar position obviously raises problems of access for initial placement and any subsequent maintenance, power generation and sky coverage (<50%). For any lunar site the velocity change required for descent to the surface (>2 km/s) additional to that needed for initial lunar transfer would be very demanding in fuel – see Figure 12, for example from Ref D4 which describes some of the steps which will be necessary to develop the infrastructure required for an international moon-based programme in the next century.

8. Conclusion

Table VII summarises the advantages and drawbacks of the various possible locations described in this paper for an EDISON-class mission. LEO and GEO are rejected on the grounds of high thermal input and radiation background respectively (though a small precursor mission could sensibly adopt a sun synchronous LEO). Elliptical orbits, similar to ISO/FIRST are possible subject to further assessments of the equilibrium temperature particularly as a function of perigee height. High circular orbits(e.g. 100,000 km) offer clear advantages to performance but at the expense of very large launch/boost requirements which limit the attainable on-orbit mass and hence the size of the telescope. An orbit at the Sun-Earth L2 libration point has all the advantages of a high circular orbit plus improved instantaneous sky

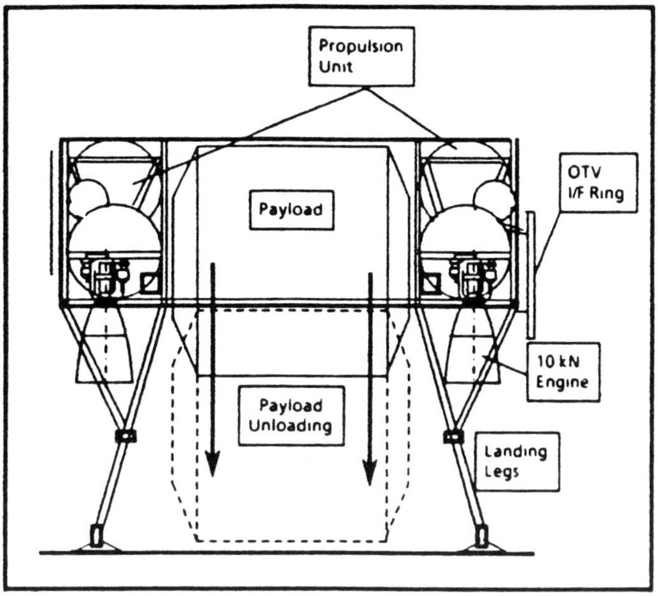

Fig. 12. Conceptual lunar landing vehicle.

coverage. The earth-to-spacecraft range (1.5 x 10^6 km) is expected to be acceptable for data retrieval at rates up to 1Mbps.

It is recommended that further work should address the mission and system design of a radiatively-cooled telescope in such an orbit with emphasis on payload mass/size and temperature optimisation.

References

CRYOGENICALLY COOLED TELESCOPES

A1 Cesarsky, C.J. and Kessler, M.F.: 1990. "The Infrared Space Observatory", p. 205-214 in *Proceedings of the 123rd IAU Colloquium, "Observatories in Earth Orbit and Beyond"*, edited by Y. Kondo. Kluwer Academic Publishers.

A2 ESA Astronomy Working Group 1987. ISO 24 hour Orbit. ESA/SPC (87) 14 plus Attachment ASTRO (87)7.

A3 Mather, J.C. *et al.*: 1990. "Early Results from the Cosmic Background Explorer (COBE)", p. 3-8 in *"Observatories in Earth Orbit and Beyond"* (see Ref.A1).

A4 Werner, M.W. *et al.*: 1989. "SIRTF in High Earth Orbit" p. 387– 400 in *Proceedings of the 3rd Infrared Detector Technology Workshop*. NASA TM-102209.

A5 Fazio, G.G. and Eisenhardt, P.: 1990. "The Space Infrared Telescope Facility (SIRTF)" p. 193-203 in *"Observatories in Earth Orbit and Beyond"*. (See Ref A1).

A6 ESA 1991 STEP : Satellite Test of the Equivalence Principle. Assessment Study Report SC1(91)4.

RADIATIVELY COOLED (AMBIENT TEMPERATURE) TELESCOPES

B1 Frisk, U.: 1990. "FIRST – Far Infrared and Submillimetre Space Telescope" p. 37-45 in *ESA SP-314. Proceedings of the 29th Liege Astrophysical Colloquium "From Ground Based to Spaceborne Sub-mm Astronomy"*.

B2 Wilson, R.: 1990. "LDR – A Submillimetre 'Great' Observatory." p. 215-219 in *ESA SP-314* (See Ref. B1)
B3 Melnick, G.J. *et al.*: 1990. "The Submillimetre Wave Astronomy Satellite (SWAS)" p. 251-256 in *ESA SP-314* (See Ref.B1)
B4 Phillips, T.G.: 1990. "A Submillimetre Mission for the 1990's" p. 221-228 in *ESA SP-314* (See Ref B1)
B5 Nordh, L. *et al.*: "MOSES – A proposed Swedish Submillimetre Satellite for Combined Astronomical and Atmospheric Observations" p. 387-389 in *ESA SP-314* (See Ref B1).
B6 Strukov, I *et al.*: 1990 "RELICT-2 Project. Scientific Programme and Structure" p. 99–104 in *ESA SP–314* (See Ref B1).
B7 Thronson, H.A. *et al.*: 1990 "EDISON : A Second Generation Infrared Space Observatory" p. 501-506 in *"Observatories in Earth Orbit and Beyond"* (See Ref A1).
B8 Davies, J.K. *et al.*: 1990. *Radiatively Cooled Telescopes: A New Direction for Infrared Astronomy.* Acta Astronautica. **25**, 233.
B9 Hawarden, T.G. *et al.*: 1989. "POIROT : A Passively Cooled Orbiting Infrared Telescope." Proposal to ESA for the next medium sized (M2) project.

ORBIT TRADE-OFFS
C1 Roth, E.A.: 1984. "Mission Analysis Aspects." p. 113–116 in *ESA SP 226. Kilometric Optical Arrays in Space.*
C2 Taylor, B.G.: 1990. "Relative Merits of Low Earth, Eccentric, Geosynchronous and Interplanetary Orbits and Sites in Space." p. 377–379 in *"Observatories in Earth Orbit and Beyond"* (See Ref A1).
C3 Farquhar, R.W. and Dunham, D.W.: 1990. "Use of Libration – Point Orbits for Space Observatories." p. 391–395 in *"Observatories in Earth Orbit and Beyond"* (see Ref A1).

LUNAR BASED ASTRONOMY
D1 Smith, H.J.: 1990. "Lunar Based Astronomy." p. 365–375 in *"Observatories in Earth Orbit and Beyond"* (See Ref A1).
D2 Mumma, M.J. and Smith, H.J. (Ed): 1990. *Astrophysics from the Moon.* AIP Conference Proceedings **207**.
D3 NASA 1988. *Proceedings of Workshop on Future Astronomical Observatories on the Moon.* NASA Conference Publication **2489**.
D4 Parkinson, R.C., Conchie P.J. and Hempsell, C.M. *Potential Evolution of an International Moon Based Programme.* IAA-90-641.

LAUNCHERS
E1 Krause, R.B.: 1990. "United States Launch Vehicle Systems." p. 325 - 332 in *"Observatories in Earth Orbit and Beyond"*. (See Ref A1).
E2 Feustel-Beuchel, J. and Wamstekeer, W. ESA's Space Transportation System. p. 333-338 in *"Observatories in Earth Orbit and Beyond"*. (See Ref A1).

OPTICAL DESIGN AND TESTING OF A FAST, LARGE APERTURE, INFRARED SPACE TELESCOPE

C.M. HUMPHRIES, Y. NEVO, E. ETTEDGUI-ATAD and J.W. HARRIS
Royal Observatory, Blackford Hill, Edinburgh EH9 3HJ, U.K.

Abstract. An optical design study for a next generation infrared space telescope has been performed. The concept is that of a passively cooled telescope of minium aperture 2.5 metre with an F/1.2 primary and wavelength coverage from $\lambda = 2$ to at least 40 μm, and possibly to 100 μm. Compactness, low thermal emission from the optics and structure, diffraction limited imaging at $\lambda = 2$ μm, and sensitivity to misalignment aberrations and manufacturing errors were the main considerations for this study. Ray tracing results are presented showing the characteristics of the various designs considered. A preliminary investigation of stray light properties is also given. Special emphasis has been placed on the testing of such a fast primary, and optical systems using a lateral shearing interferometer are described for testing both the primary and the primary/secondary combination.

1. Introduction

The proposed design of the passively cooled EDISON infrared space telescope (Thronson *et al.*, 1990) is such that it relies on radiative cooling to reach temperatures less than 60K, with additional cooling for the detector components provided by a mechanical refrigerator, or by a hybrid refrigerator/cryogenic system, to approximately 10K. This would enable the spacecraft to function at these temperatures for an extended lifetime without the need for cryogenic liquid replenishment. With a fast (F/1.2), large aperture (≥ 2.5m) primary mirror, the aspect ratio of the telescope tube is well suited to give efficient radiative cooling and reduce thermal gradients along the tube. At the same time it gives a substantial increase in aperture over other infrared space telescopes, both existing and planned (e.g. IRAS 0.6m, ISO 0.6m, SIRTF 1.0m).

With spectral coverage from $\lambda = 2$ to 40+ μm, the design objective was to achieve diffraction limited imaging at the shortest wavelength of this range, corresponding to images with FWHM ≤ 0.2 arcsec for a telescope aperture of 2.5 metre. The field of view over which imaging performances were investigated was set by (a) taking the focal aperture ratio of the secondary beam to be F/15, (b) assuming that the detectors are 256×256 arrays with pixel size proportional to wavelength, (c) combining the diffraction sizes of the images at the long wavelength end of the range considered with image sampling considerations. With a pyramid feed to four separate instruments, the central 2 arcmin field is used for guiding and the total acceptable FOV within the above restrictions was found to be about 6 arcmin radius centred on the telescope optical axis (Figure 1). The four representative fields A, B.... in Figure 1 are each divided into an inner portion, where the design requirement is for short wavelength diffraction limited imaging,

Space Science Reviews **61:** 211–232, 1992.

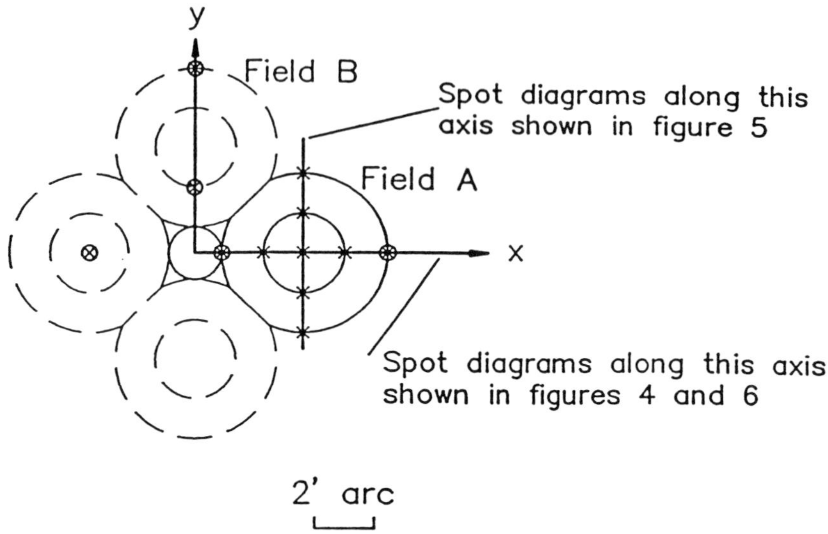

Fig. 1. Representative fields of view investigated for four instruments with a pyramid feed (crosses indicate the locations where performances were investigated; circled crosses indicate field positions used to optimise the off-axis designs).

and an outer portion where larger field-dependent aberrations can be tolerated at longer wavelengths. Further details of the imaging performances are discussed in Section 3.2.

2. Optical Configurations

Two types of system have been studied: (a) on-axis systems having a central obscuration (obscuration ratio $\varepsilon = 0.1$), with an F/15 secondary beam and a back focal distance approximately 1 metre behind the pole of the primary, (b) unobscured systems, also with an F/15 secondary beam and a similar back focal distance. In each of the cases considered the secondary is taken to be the aperture stop. The choice of F/15 was made as a compromise between (1) the focal plane scale, and therefore the physical dimensions for a given field of view; and (2) the size of secondary mirror needed to obtain a particular focal plane scale. The first of these considerations influences the size of optical components in instruments around the focal plane, while the second determines the obscuration ratio of the telescope and affects also the radiative emission received by the detectors as well as the power requirements for an active secondary control system.

2.1. Systems with Central Obscuration

Design data for classical Cassegrain and Ritchey-Chrétien systems with an F/1.2 primary are given in Table 1 and shown as a layout in Figure 2(a). In each case, the conic surfaces were adjusted to optimise the image performances at five locations

TABLE I
Optimised design data for classical Cassegrain (CC) and Ritchey-Chrétien (RC) systems

System Parameters	CC	RC
Primary radius of curvature	-6000.00	-6000.00
Conic Constant (K_1)	-1.00000	-1.00143
Axial separation	-2700.00	-2700.00
Secondary radius of curvature	-652.17	-652.17
Conic Constant (K_2)	-1.37812	-1.39644
Image distance from secondary	3750.12	3750.12
Image surface radius of curvature	-314.94	-312.36

TABLE II
Optimised design data for off-axis systems: Ritchey-Chrétien, aplanatic Gregorian, and folded Ritchey-Chrétien

System Parameters	off-axis RC	off-axis AG	off-axis FRC
Primary radius of curvature	-6000.00	-6000.00	-6000.00
Conic Constant (K_1)	-1.00146	-0.99866	-1.00577
Entrance pupil decentre for axisymmetric system	1650.00	1250.00	3750.00
Axial separation	-2700.00	-3347.20	-5420.00
Secondary radius of curvature	-652.17	642.97	1380.95
Conic Constant (K_2)	-1.39676	-0.73484	-2.00797
Image distance from secondary	3750.28	4340.17	3624.93
Image surface radius of curvature	-317.58	342.14	-592.58

across a diameter (x-direction) of field A in Figure 1 for a spherical focal surface. The weights given to these field positions were in the ratio 0.6:0.875:1.0:0.875:0.6.

2.2. Decentred Pupil Systems without Obscuration

Layouts for an off-axis Ritchey-Chrétien, an off-axis aplanatic Gregorian, and an off-axis folded Ritchey-Chrétien are shown in Figures 2(b)-(d). Although the term 'off-axis' is used here for convenience to describe configurations 2(b) and 2(c), they are more properly considered as F/1.2 portions of axisymmetric systems having an F/0.6 parent primary but with a decentred pupil. Similarly, the primary mirror in the folded system of configuration 2(d) is itself only F/2.4 but is a decentred pupil of an axisymmetric F/1.2 primary.

Optimised design data are given in Table 2. The optimisations were performed

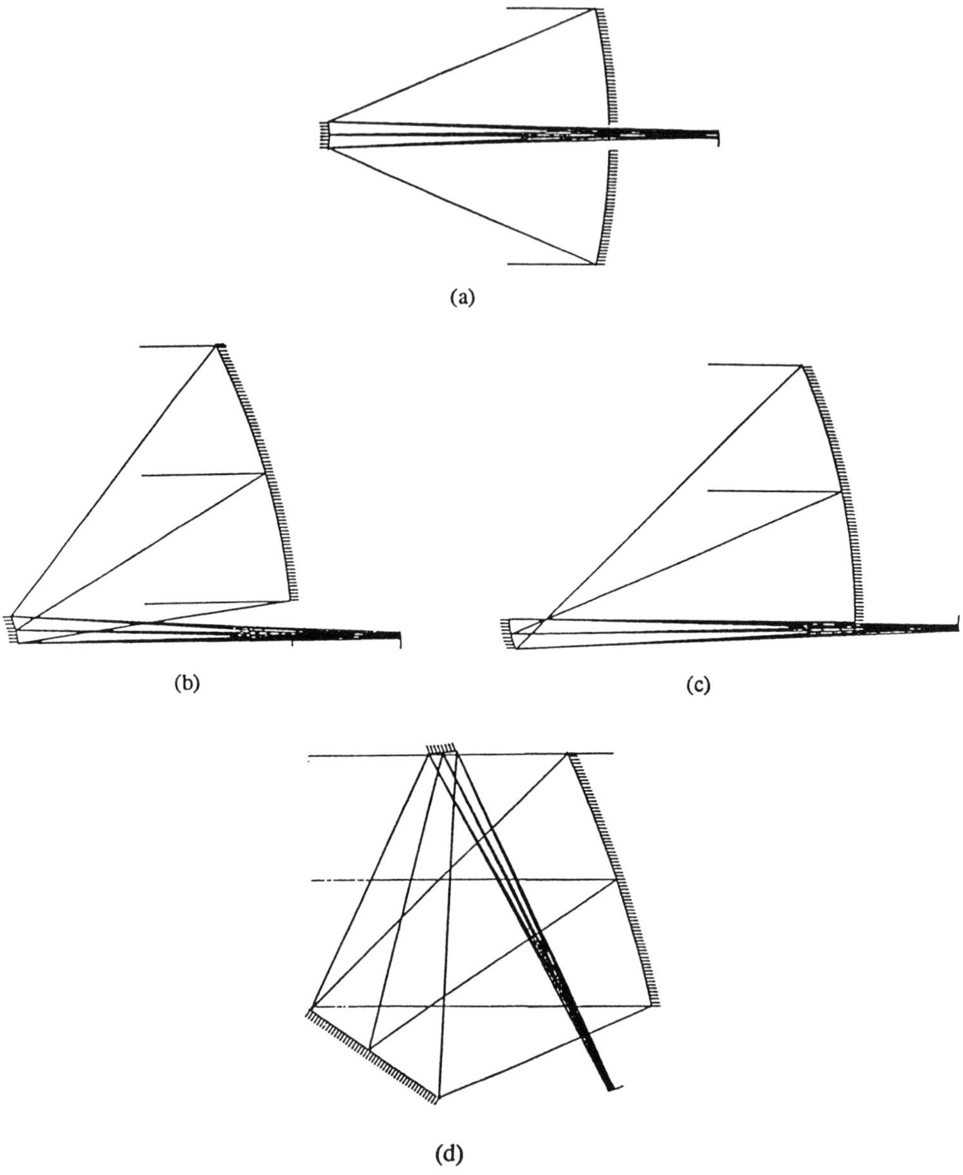

Fig. 2. Telescope configuration layouts (a) on-axis Cassegrain/Ritchey-Chrétien (b) off-axis Ritchey-Chrétien (c) off-axis Gregorian (d) off-axis folded Ritchey-Chrétien.

over five positions, equally weighted, covering the x and y directions of fields A, B... in Figure 1, again allowing spherical focal surfaces. The main characteristics of the designs are summarised as follows: (a) the off-axis R-C is the most compact; (b) an off-axis Gregorian is longer but has an accessible intermediate focus as well as a real exit pupil; the longer tube will lead to a slightly higher tube temperature by restricting radiative cooling; (c) the folded R-C has lower sensitivity to misalignment, as described later, but is quite compact despite the F/2.4 primary; however, it requires the addition of a 1.5 metre diameter flat and it sacrifices telescope aperture for a given launcher diameter.

2.3. Instrument Optical Feeds

For the on-axis Cassegrain/R-C configurations, a pyramid reflector positioned before the telescope focus reflects light to each of four instrumentation packages, as, for example, with HST and ISO. Other geometric arrangements could be accommodated by changing the distribution and angles of the pyramid faces. The advantage of the pyramid reflector is that it has no moving parts and thus has reliability and no associated heat sources (such as motors); also, it uses all of the light being viewed within the passband of a single instrument. The principal disadvantages are that a single object cannot be viewed simultaneously by the four instruments, light of wavelengths beyond the passband of each instrument is not utilised for any other purpose, and the total FOV is larger than would otherwise be required. Other solutions involving dichroic beam splitters or movable mirrors, though feasible and possibly better suited to particular requirements of volume and geometry, are not considered further here. Thus, in the present study a pyramid reflector is envisaged also for use with the off-axis telescope designs so as to obtain representative fields of view that can be directly compared in imaging performance with those of the on-axis systems.

3. Imaging Performances

3.1. On-axis Systems

Imaging performances have been investigated by ray tracing at specific locations within one of the instrumental package FOVs of Figure 1 (since symmetry produces identical results for the others). Three types of focal surface have been considered: (i) a spherical surface; (ii) a tilted plane; and (iii) a plane, orthogonal to the optical axis. At each location the RMS wavefront error was calculated and the results for the spherical focal surface are presented in Figure 3 (top). Variables for the optimisation were the conic constants K_1 and K_2 as well as the back focal distance and the curvature of the focal surface. The diffraction limits shown are the permissible $\lambda/14$ RMS wavefront height errors appropriate to the telescope at the stated wavelength.

Figure 3 shows that, with spherical focal surfaces, $\lambda = 2\ \mu$m diffraction limited imaging is obtained out to a radius of $\sim$ 6 arcmin for both the Cassegrain and the Ritchey-Chrétien. As expected, the R-C has the better imaging at a given field radius but the difference is only marginal for the conditions assumed. When the curved focal surfaces are approximated by plane surfaces set orthogonally to the respective optical axes, then the results shown in the lower part of Figure 3 apply. Here, the axial position of each focal surface has been adjusted so that aberrations

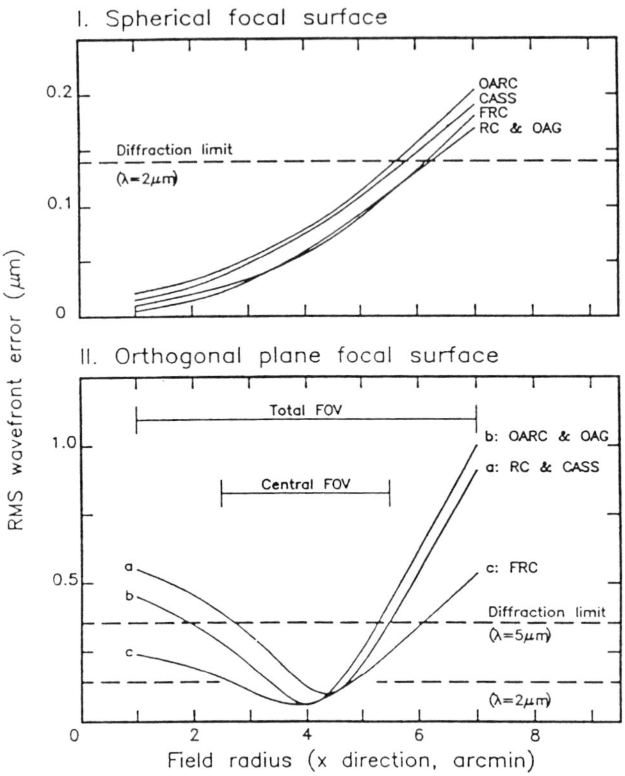

Fig. 3. RMS wavefront errors of (a) classical Cassegrain (CC) and Ritchey-Chrétien (RC), (b) off-axis Ritchey-Chrétien (OARC) and off-axis aplanatic Gregorian (OAG), (c) folded Ritchey-Chrétien (FRC). In the top set of curves, spherical focal surfaces have been assumed; in the lower, the focal surfaces are planar orthogonal.

are minimised at a field radius of approximately 4 arcmin from the telescope optical axis.

That the Cassegrain and RC configurations have such similar RMS wavefront errors in curve (a) of the lower portion of Figure 3 is a consequence of (1) having identical Petzval curvature, and (2) the mismatch of that curvature with an orthogonal plane focal surface. With the tilted plane focal surface, the FOV from 1.5 to 6 arcmin radius is diffraction limited at $\lambda = 2\ \mu$m for both the Cassegrain and the Ritchey-Chrétien.

The planar orthogonal focal surface is very much the worst case imaging situation and has been included only to cover the possibility of direct focal plane detection by, for example, a flat array device. In all other cases, where re-imaging is used, additional optical surfaces will be available for wavefront processing so that the original design imaging characteristics (curved focal surface) may be recovered even if the detector is then flat.

3.2. OFF-AXIS SYSTEMS

For the off-axis telescopes the same FOV layout (Figure 1) has been kept. Although the actual distribution of fields may differ, based on practical considerations as well as the number and type of instrumentation packages chosen, the field radius over which images are sampled is not likely to change significantly.

For the limited field radii considered, the RMS wavefront errors of the off-axis (decentred axisymmetric) configurations show only small differences from the on-axis systems when spherical focal surfaces are assumed (Figure 3). With the exception of the folded Ritchey-Chrétien, which shows an advantage resulting from the shallower parent primary, this applies also when the focal surfaces are planar orthogonal. It is apparent from Figure 3 that whereas with spherical focal surfaces all configurations give diffraction limited imaging at $\lambda = 2$ μm over most of the field considered (to x=7 arcmin), diffraction limited fields using planar focal surfaces are restricted to a central 1 arcmin diameter (2.5 arcmin diameter for the FRC) at $\lambda = 2$ μm. The central FOV regions shown for each field in Figure 1 are those over which $\lambda = 5$ μm diffraction limited imaging is obtained by direct focal plane detection at a planar focal surface.

Spot diagrams showing the imaging characteristics of all the systems studied are presented in Figures 4, 5 and 6. Figure 4 shows these for a diameter across the x axis of Field A (cf. Figure 1) using spherical surfaces; Figures 5 and 6 show the spot diagrams for x and y diameters of field A with an orthogonal planar focal surface for each system.

4. Sensitivities to Manufacturing Errors, Misalignment, and Temperature Gradients

4.1. MANUFACTURING ERRORS

For each of the configurations studied, errors in radius of curvature of either the primary or the secondary may be compensated satisfactorily by changing the mirror separation accordingly. For example, a 0.2% error in the primary radius ($\pm$12mm) for the off-axis Ritchey-Chrétien is fully compensated by a shift in the axial position of the secondary of $\pm$6mm, keeping the RMS wavefront error within the nominal value of $\sim$ 0.05 μm at the centre of the field. Similar behaviour is exhibited by each of the other systems.

Low frequency surface errors (either from manufacturing errors in the mirror aspheric coefficients, producing spherical aberration due to the mismatch between K_1 and K_2, or from aberrations such as astigmatism) need to be kept below $\lambda/8\sqrt{2}$ peak-to-valley for each of the primary and secondary mirrors if the performance is to remain diffraction limited. Thus, at $\lambda = 2$ μm the permissible p-v error in the mirror surfaces is 0.18 μm and if, for the primary, this was all due to an error in K_1 then ΔK_1 has to be less than 1.3×10^{-4}.

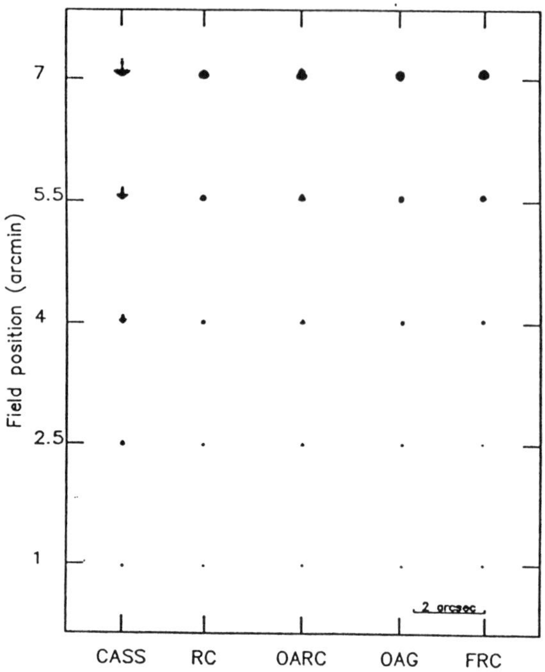

Fig. 4. Spot diagrams across the x axis of field A in Figure 1 for the optical systems studied, with spherical focal surfaces.

4.2. Misalignment – Decentre and Tilt

Sensitivities of the Ritchey-Chrétien systems to wavefront errors caused by decentre and tilt of the secondary mirror were investigated by ray tracing where (a) only the focal surface axial position was optimised, and (b) both the secondary mirror and the focal surface axial positions were allowed to vary until optimised after introducing the misalignment. The results are displayed in Figures 7 and 8.

For the off-axis R-C (curve 1a in each figure) the wavefront error due to decentre or tilt can be reduced by adjusting the axial position of the secondary (curve 1b). For the on-axis systems this is not necessary since cases 2a and 2b are identical, or nearly so, in Figures 7 and 8. It should be noted that for the off-axis systems, decentre, tilt and despace are referred to the optical axis of the parent surface for the primary (so that adjusting the axial position of the secondary involves movement along the latter axis and not along the centre-to-centre axis from the off-axis primary segment to the secondary). The off-axis Gregorian sensitivities are similar to those of the off-axis Ritchey-Chrétien. For the folded Ritchey-Chrétien the sensitivity of wavefront error to decentre and tilt is much less than for the other off-axis systems. Again this is due to the slower effective primary focal aperture ratio (2.4 instead of 1.2).

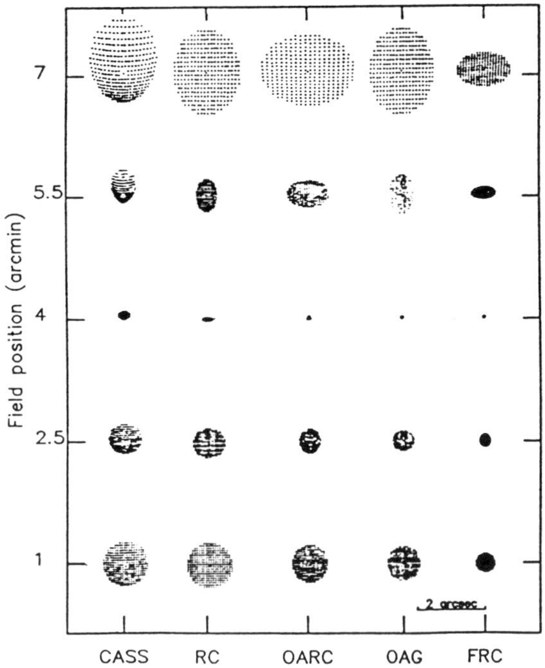

Fig. 5. Spot diagrams across the x axis of field A in Figure 1 assuming planar orthogonal focal surfaces.

4.3. Misalignment – Despace

Figure 9 shows the sensitivities due to despace errors. Again, the off-axis Ritchey-Chrétien is more sensitive than the on-axis system whereas the folded Ritchey-Chrétien has a sensitivity close to that of the on-axis system. The on-axis system can tolerate a despace error of ~ 0.5mm, and the FRC a despace error of ~ 0.12mm, while the off-axis Ritchey-Chrétien and Gregorian axial mirror positions must be within 0.03mm of the nominal spacing requirement for these systems to remain diffraction limited at $\lambda = 2\ \mu$m.

4.4. Temperature Gradients

Ray tracing simulations were performed for the on-axis RC system to predict the effects of (a) cooling the system from room temperature to 77K, assuming no temperature gradients after thermal equilibrium, and (b) a temperature gradient of 10K between the primary and secondary mirrors. The material chosen for the simulations was aluminium alloy and the temperature sensitivities of this and other candidate materials are shown in Figure 10.

Cooling the telescope from 293K to 77K produces a decrease in the intermirror separation of 0.4%, together with decreases in the primary and secondary radii of curvature by the same percentage. Apart from the change in dimensions and a

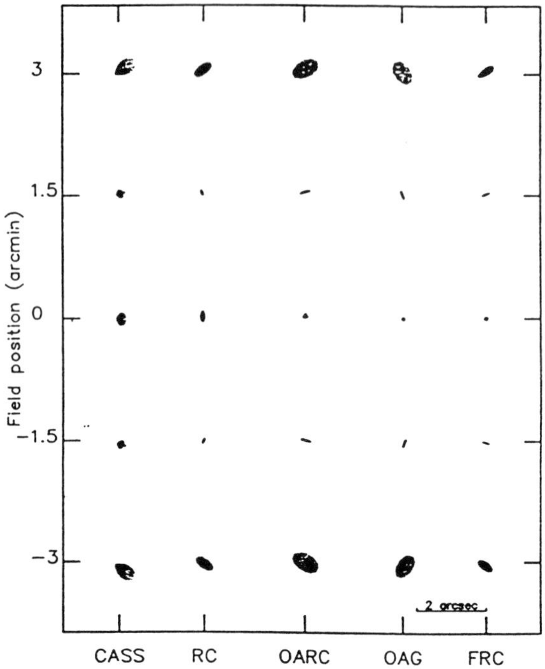

Fig. 6. Spot diagrams across a diameter of field A in Figure 1 parallel to the y axis, assuming planar orthogonal focal surfaces.

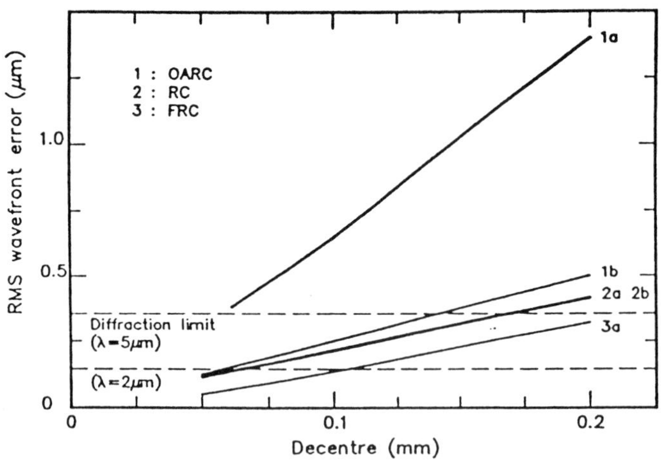

Fig. 7. Effect of secondary mirror decentre on the RMS wavefront error (a) with only the focal surface axial position optimised (labelled a); (b) with secondary mirror and focal surface axial positions optimised (labelled b).

change in the focal plane plate scale (corresponding to a change in focal aperture ratio from F/15 to F/14.94), the optical imaging performance suffers no degradation. This assumes that the material used for the telescope structure is of the same

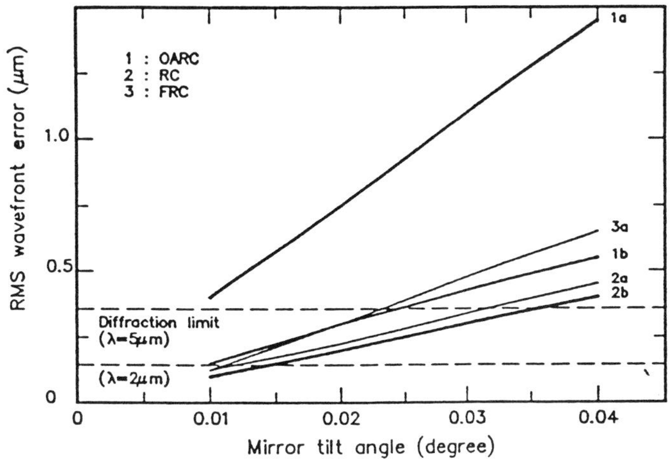

Fig. 8. Effect of secondary mirror tilt on RMS wavefront error; cases (a) and (b) are as for Figure 7.

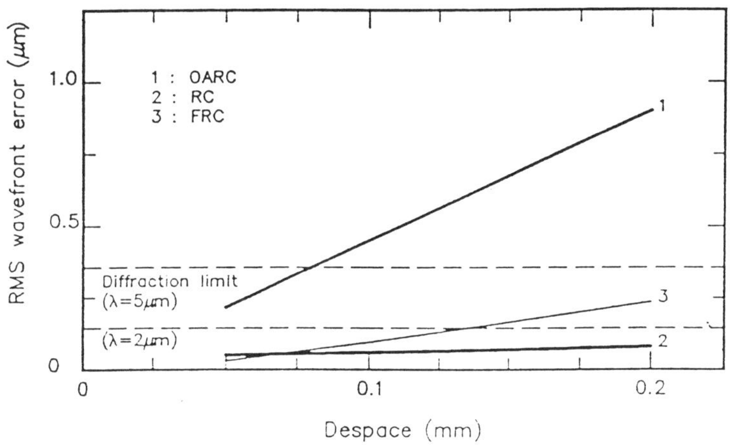

Fig. 9. Effect of secondary mirror despace on RMS wavefront error with focal surface axial position optimised.

composition as that of the mirrors themselves, which is the optimum design choice for preserving the image quality when cooled in space. Such an approach has been used successfully in the design, *inter alia*, of cryogenically cooled instruments for ground based telescopes and simplifies both the mechanical design requirements and the optical testing of the telescope.

For a temperature gradient of 10K between the primary and secondary, the differential effect on the secondary radius of curvature ($\Delta R_s = -6.5$ μm) is so small that, again, the optical image quality remains diffraction limited at $\lambda = 2$ μm in each FOV when using spherical focal surfaces. If differential thermal effects between the telescope structure and the optics are allowed, with adjustments to re-optimise the focal surface position and possibly the inter-mirror separation

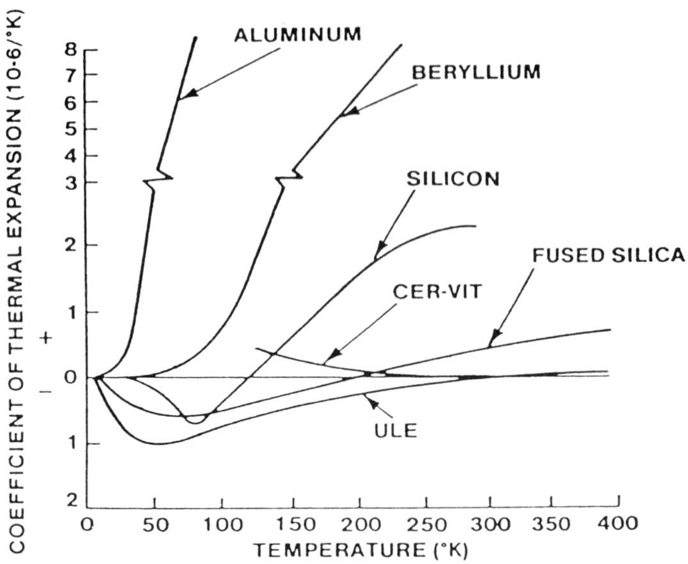

Fig. 10. Thermal expansion coefficients of mirror materials [from Paquin and Goggin (1972), as presented by Yoder (1986)].

TABLE III
Summary of physical properties of mirror materials

	A1-alloy (6061-T6)	Beryllium	Fused Silica
Density (g cm^{-3})	2.71	1.85	2.20
$\alpha_{25} \times 10^6$ (°C)$^{-1}$	23.9	11.2	0.56
Stiffness to weight ratio (10^4 N m g^{-1})	2.54	16.4	3.3

at temperatures other than those of the nominal design, other materials such as beryllium or fused silica may be considered for the optics. The densities, thermal coefficients at 25°C, and stiffness to weight ratios of these and of aluminium alloy (6061-T6) are compared in Table 3.

5. Stray Radiation Considerations

5.1. STRAY LIGHT

Stray light properties of the telescope can, for convenience here, be divided into four categories. These will be referred to as (a) thermal emission, (b) low intensity image structure, (c) background radiation, and (d) skylight. Thermal emission is the radiation that is emitted by surfaces within the beam or scattered by these surfaces into the beam from other emitting surfaces. Microroughness and particulate contamination of mirror surfaces scatter energy from the centre of images out into the wings while obstructions diffract energy also to the wings. Outside the telescope beam, as seen from the telescope focal plane, there are surfaces that radiate thermal emission to the telescope focal plane and that may also scatter significant amounts of radiation to the telescope focal plane from the Sun, Earth or Moon. This background radiation may be a problem when detectors are used at the telescope focal plane, otherwise it can be controlled within the instruments as is the case with ground based telescopes. Skylight is integrated radiation falling from the sky directly on to the telescope focal plane. Again it is likely to be a problem only for telescope focal plane detectors.

5.2. THERMAL EMISSION AND RADIATION SCATTERED INTO THE BEAM

Principal contributions to the thermal emission arise from self emission of the primary and secondary mirrors, and emission from the tube interior scattered by the mirror surfaces into the beam. In the case of centrally obstructed systems there will also be emission from the secondary support vanes and from the hole in the beam seen by reflection in the secondary. If sufficient radiation from the Sun, Earth or Moon reaches surfaces seen by the mirrors, then radiation scattered by the mirrors into the beam from these sources may also need to be considered.

Blackbody emission from mirrors at a temperature of 40K would equal that from the celestial background at $\lambda \simeq 40\ \mu$m. For real mirror surfaces, with lower emissivities, higher temperatures are acceptable. Simple models for the telescope cooling indicate that the temperature of the tube interior may be as low as 60K when cooling is from the tube interior alone but, to reach lower temperatures, radiators connected to the shields have to be rather effective. For a temperature of 60K, blackbody radiation equals that from the celestial background at about 25 μm wavelength and exceeds it by a factor of about 40 at a wavelength of 40 μm. Even with coefficients of emissivity of low as 0.01 the combined self emissions of two mirrors at this temperature will just reach the celestial background at 40 μm. In the respect, the off-axis folded Ritchey-Chrétien telescope gives higher emission because of the extra mirror. On the other hand, the on-axis systems may have an advantage if the secondary is cooled; here, the secondary is in a favourable position for radiating to space and, with non-conductive supports, simple considerations indicate that the mirror temperature will be reduced to between a half to one-third of the tube temperature which would make the secondary self emission unimportant. The off-axis Ritchey-Chrétien and Gregorian configurations also have their secondary mirrors near the open end of the tube and so could be radiatively cooled, but not so effectively. This is because the off-centred secondary sees a smaller solid angle of sky.

For the centrally obscured systems, the hole in the primary seen by reflection in the secondary is highly emissive, assuming temperatures similar to that of the tube, but it comprises typically only 1% of the area of the beam. This is equivalent to an addition to the emission from the secondary, raising its apparent coefficient of emissivity by 0.01 which is enough to offset any gain in cooling if the actual coefficient of emissivity is as low as 0.01. Of course, measures can be taken to ensure that the sky is seen, either directly or indirectly by reflection in the primary, rather than the hole in the secondary. The secondary support structure is also seen within the beam and may comprise, say, 0.5% of the beam area. Again it is convenient to attribute the emission from these supports to the secondary, which raises the apparent emissivity by up to 0.005 depending on any measures taken to reduce this figure as well as the temperature of the secondary support relative to the tube temperature.

Scatter at the mirror surfaces depends on the aims and method of manufacture which may be different for the different telescope configurations. To obtain some idea of the level of scatter it is necessary to assume either the total integrated scatter (TIS) or the bidirectional reflectance distribution function (BRDF). We adopt the nominal BRDF mirror scattering data given by Evans and Breault (1984). This has properties intermediate between diamond turned and polished surfaces, and gives a TIS of about $4 \times 10^{-3}\%$ at 10 μm which corresponds to a surface roughness of 5 nm RMS. At $\lambda = 40$ μm the TIS is smaller by a factor of 16 and at this level of scatter, thermal emission from the tube at 60K, scattered by the mirrors into the beam, will be negligible.

Particulate contamination of the mirror surfaces can increase scatter to the point where it becomes significant. We use the preducted BRDF data given by Young (1976). At 10 μm wavelength and cleanliness level 500 (dirty by optical standards), the TIS is about 0.5%. For two mirrors this gives a TIS of 1% but this will be less at 40 μm. Also, since small angle scatter is strong, radiation is scattered into the beam predominantly from the sky rather than the tube. Thermal radiation from the tube scattered into the beam is unlikely to be a problem if the mirrors can be cleaner than level 500.

5.3. Low Intensity Image Structure

Energy scattered out to the wings of an image, or diffracted there, can reduce the ability to discriminate faint objects near to bright ones. The centrally obstructed systems differ from the others in the effect of the obscuration and the support structure. The former reduces the central diffraction maximum and transfers energy to the ring structure but the effect is very small for an obscuration ratio of 0.1. Diffraction spikes produced by the mirror spider assembly may be a considerable nuisance as the energy is concentrated in specific directions and falls off with the inverse of the square of the radius rather than the inverse of the cube of the radius as is the case with the smoothed out ring structure of the image. Scatter by microroughness and zonal errors is common to all the systems but is possibly higher for the off-axis systems since unconventional technology is required for their manufacture. The folded Ritchey-Chrétien will have additional scatter because of the extra mirror but as this is a flat, the scatter of this mirror can be kept low. Scatter by particulates is particularly strong at short wavelengths for small angle scatter and in this respect the folded system is clearly at a disadvantage with the extra mirror.

5.4. Background Radiation

Radiation outside the beam, when viewed from the telescope focal plane and which does not include the component from the sky seen directly, can be large in comparison to the radiation within the beam because the solid angle subtended is nearly a hemisphere. However, a focal plane detector will receive less radiation from a surrounding thermal background at 60K than from the celestial background seen through the telescope if the wavelength is shorter than 10 μm. In order not to add to the noise from the celestial background by more than 10% at $\lambda = 40$ μm, the surfaces surrounding focal plane detectors and external to any filters would have to be at a temperature of 20K or less. This implies the use of a baffle cooled below the tube temperature and positioned to the rear of the primary or in the case of the folded Ritchey-Chrétien, forward of the focus but outside the ray envelope between the primary and flat in Figure 2. Such baffles are not very effective at restricting solid angle. For a field of 10 arcminutes, the solid angle is restricted to approximately 5 times that of the beam for the on-axis configuration and is worst for the folded configuration where the factor is about 10. Note also that when baffled, the detector field of view outside the beam is wholly filled by radiation from the tube for the folded Ritchey-Chrétien but no tube radiation is directly seen with the on-axis system. Tube radiation seen by detectors in the other off-axis configurations varies between these two extremes according to the field position. Consequently, when cooled baffles are provided, the long wavelength limit set by noise from the thermal background at 60K is raised to $\sim$ 16 μm for the folded Ritchey-Chrétien, to $>$ 40 μm for the on-axis systems, and between these limits for the other configurations.

Where instrumentation re-images the telescope focal plane, background radiation and sky light is suppressed by the use of a Lyot stop and other cold baffles. The on-axis configuration has a clear advantage over the other configurations because radiation close to the beam is from the sky rather than the tube. For the folded Ritchey-Chrétien, radiation from the tube at 60K from an annulus around the secondary will contribute 10% to the noise at 40 μm wavelength if the stop allows the effective area of the annulus to equal 0.5% of the area of the secondary.

5.5. Skylight

It is normal to use sky baffles in visible light telescopes and with cooled space telescopes used at both short and long wavelengths to prevent radiation from the sky passing directly to the telescope focal plane detectors. Sufficiently cold baffles for use at mid-infrared wavelengths must be placed near, or form part of, the instrumentation package and, as already noted, they also restrict background radiation reaching the detectors. Despite this, the on-axis systems, when baffled in this way, still pass enough sky radiation to more than double the noise relative to that from the celestial background seen through the telescope. On the other hand, the baffled folded Ritchey-Chrétien can receive skylight only by scattering from the tube and so the increase in noise should be negligible. The other off-axis systems are intermediate in behaviour and field dependent. Instrumentation that follows the telescope focal plane and is adequately baffled against radiation from the tube should present less of a problem in baffling against sky radiation.

6. Optical Testing

6.1. Testing Philosophy

If the material of the optical mounting structure is the same as that of the optics, for example with metal mirrors, then the linear reduction of the primary radius of curvature as the temperature is reduced to cryogenic temperatures is matched by a corresponding contraction of the support structure. Thus the optical system remains in focus. This is a first order description only, however, and in practice it is found that additional changes in mirror surface figure may occur at cryogenic temperatures. For example, the IRAS primary mirror was specially figured to compensate for surface errors at low temperature (Harned *et al.*, 1980). In the present case, it is expected that testing in a liquid nitrogen cooled cryogenic tank would be sufficient to determine and correct for these errors since almost all of the figure change from room temperature will already have occurred at 77° K. Thus, even with metal mirrors optical testing at cryogenic temperatures is required, as also is the case if other materials are used for the optics.

The tests in the following Sections are described for use at room temperature. At cryogenic temperatures for an on-axis primary mirror (Section 6.2), the same test is employed but the test equipment of 6.2, including the null compensator, is separated from the optics under test by a window and thus is external to the cryostat chamber. For off-axis primary mirror testing using auto-collimation (Section 6.3), the cryogenic chamber has to be large enough to include the auto-collimating flat, and the surface errors of the latter at cryogenic temperatures are calibrated separately. For testing the complete telescope at cryogenic temperatures, methods (b) and (c) of Section 6.4 can be applied, using a window to separate the test equipment from the cryogenically cooled telescope.

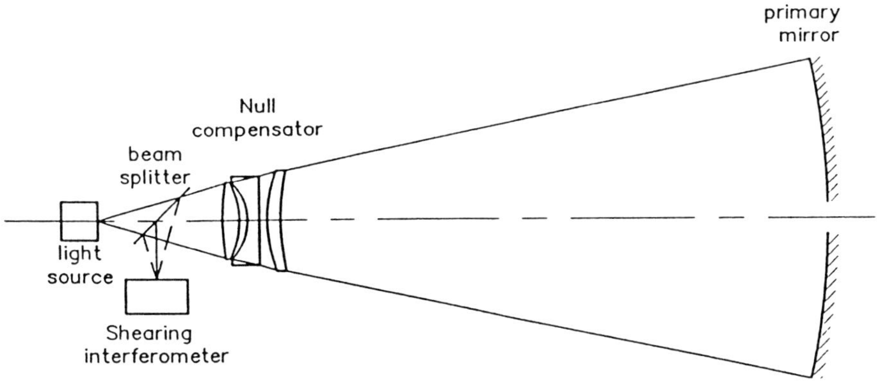

Fig. 11. Test system for on-axis primary.

6.2. ON-AXIS PRIMARY

For such a fast primary mirror, a three element refractive null lens system is required for testing the primary at centre of curvature (principally to compensate the large amount of spherical aberration – 440 waves RMS at $\lambda = 0.5$ μm for a 2.5 metre Ritchey-Chrétien primary). Figure 11 shows a triplet null corrector inserted in a double pass test configuration suitable for analysis by interferometry. The design given in Table 4 produces a residual wavefront error of 0.31λ RMS at $\lambda = 0.5$ μm. Figure 12 shows the transverse ray aberrations plotted against pupil radius for the tangential and sagittal rays. Although the null corrector itself still does not give quite the precision needed for testing a $\lambda = 2$ μm diffraction limited system, the design and manufacturing errors of the corrector can be calibrated and subtracted from the interferometric test. This seems preferable to the alternative of improving the compensation by adding a further element to the corrector and thereby increasing the complexity and cost; for the corrector shown in Figure 11 the largest element already has a diameter of 0.5m. The interferometer in Figure 11 is of the lateral shearing type; apart from insensitivity to thermal fluctuations and mechanical vibrations it also has the capability of measuring rotationally symmetric aberrations such as spherical aberration (unlike rotational shearing interferometry).

6.3. OFF-AXIS PRIMARY

Since the off-axis primaries are actually decentred portions of much faster parent mirrors, with corresponding higher spherical aberration, testing at centre of curvature as in Section 6.1 would require a considerably more complex null corrector. As an alternative method, Figure 13 shows the source located at the focus of the primary and a flat perpendicular to the optical axis used in autocollimation. Although the testing of an off-axis paraboloid would not require compensation, the off-axis R-C considered here does. In the inset to Figure 13, the null lens is a single-pass all-spherical doublet but the curvatures of the inner component are relatively large

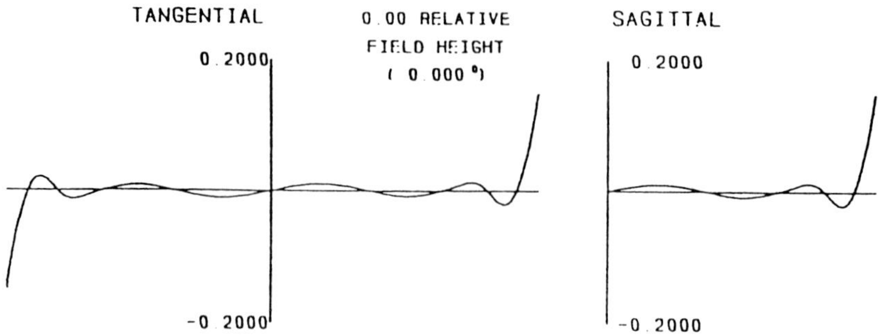

Fig. 12. Transverse ray aberrations as a function of pupil radius.

TABLE IV
Design data for Ritchey-Chrétien null compensator

Surface	Radius (mm)	Thickness (mm)	Material
Object	–	995.6	
1	2584.32	140.0	BK7
2	-553.80	50.0	
3	-456.84	100.0	BK7
4	-9782.38	70.0	
5	1303.13	100.0	BK7
6	2655.79	4266.91	
mirror	-6000.00 (K= -1.001430)		

(Table 5). The residual RMS wavefront error is 0.01λ at $\lambda = 0.5$ μm. Yet another alternative is given in Figure 14, which shows a 300mm diameter single-surface reflective compensator where the surface is aspheric with terms up to r^8 (Table 6). The residual wavefront error here is 0.1λ at $\lambda = 0.5$ μm. However, this is a design solution only since the manufacture and testing of the aspheric mirror becomes a problem in itself, thus defeating the concept of the null test.

6.4. Combined Primary/Secondary Testing

Testing of the combined primary/secondary mirror system may be accomplished by the following well known methods:

(a) Using a real or artificial point source at an infinite conjugate, and assessing the wavefront error at the focal surface of the telescope interferometrically; the advantages of lateral shearing interferometry for this purpose, and its application to optimising the performance of an astronomical telescope, have

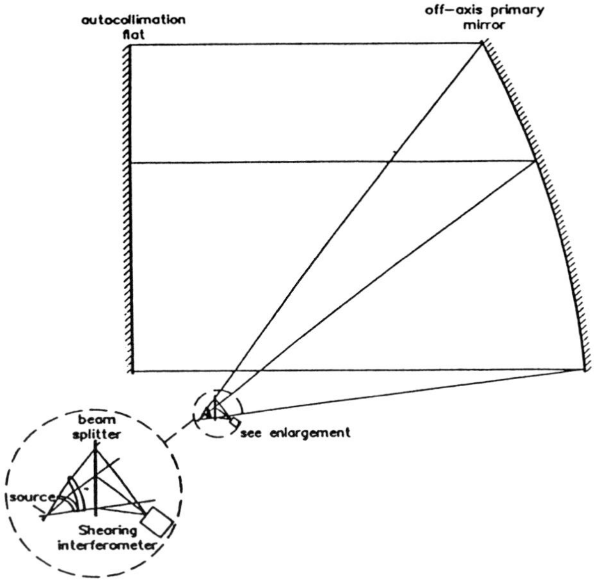

Fig. 13. Autocollimation testing of off-axis primary.

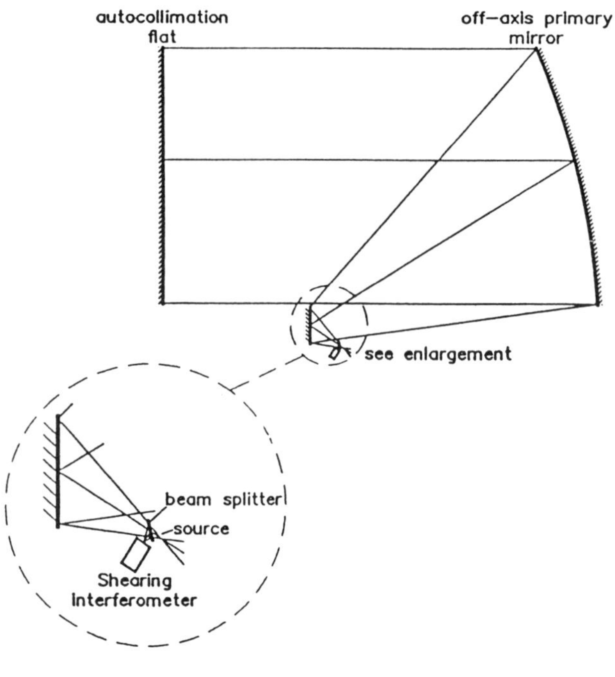

Fig. 14. Off-axis testing with reflective compensator.

been documented elsewhere (Atad *et al.*, 1990a; Atad *et al.*, 1990b).

TABLE V
Design data for off-axis null compensator

Surface	Radius (mm)	Thickness (mm)	Material
Object		65.8	
1	-42.46	10.0	SF8
2	-48.21	0.1	
3	-168.13	15.3	BK7
4	-161.37	2900.0	
Mirror	-6000.00 (K = -1.00146)	–	
Flat	–	–	

TABLE VI
Design data for off-axis reflective null compensator

Surface	Radius (mm)	Thickness
Object	–	300.0
1	2.73×10^7 K = -1.000 $A = -0.735 \times 10^{-11}$ $B = 0.195 \times 10^{-15}$ $C = -0.252 \times 10^{-21}$	-2700.0
Mirror	-6000.00 (K = -1.00146)	–
Flat	–	–

(b) Auto-collimation, with illumination by a point source at the telescope focus. However, this requires a 2.5 metre diameter flat mirror of sufficiently good optical quality ($\lambda/4$ RMS at $\lambda = 0.5\ \mu$m).

(c) With illumination as for method (b) but using a pentaprism to deflect the light perpendicularly over a small sub-aperture of the total beam. The pentaprism is mounted on an optical bench that can be translated to give coverage of the entire aperture (Espiard and Favre, 1970) and each sub-aperture is analysed interferometrically.

TABLE VII
Comparison of optical configurations

	RC/CC	OARC	OAG	FRC
Compactness/weight				
tube length	•	•	◇	•
utilisation of width	•	○	○	◇
weight	•	•	•	◇
Manufacturing				
difficulty/cost	•	○	○	◇
Testing difficulty	○	◇	◇	◇
Imaging performance	•	•	•	•
Sensitivity to				
misalignment	•	◇	◇	○
Stray radiation				
self emission	•	○	○	◇
scattered thermal				
radiation	•	•	•	•
emission from				
vanes/primary				
hole	◇	•	•	•
diffracted/scattered				
image wings	◇	•	•	○
thermal bg seen				
directly	•	○	○	◇
sky (focal plane				
instrument)	◇	○	○	•

(• = good, ○ = intermediate, ◇ = poor)

7. Summary

In Table VII we have attempted to summarise some of the main advantages and disadvantages of the optical configurations discussed in this paper. The performance categories used are only qualitative and further studies would be required to interpret these with regard to the scientific requirements for a particular mission, and to perform a cost analysis of the various proposals.

Acknowledgements

The work described in this paper was initiated by discussions with Harley Thronson of University of Wyoming and was funded in part by the Advanced Programs Branch of the Astrophysics Division of NASA. The contribution by Y. Nevo was performed during a period of sabbatical leave from Rafael in Haifa, Israel.

References

Atad, E., Harris, J.W. and Humphries, C.M.: 1990a, *"Combined Interferometry and Ray Tracing Analysis to Optimise the Alignment of Astronomical Telescopes"*, Proc. SPIE, **1271**, 340.

Atad, E., Harris, J.W., Humphries, C.M. and Salter, V.: 1990b, *"Lateral Shearing Interferometry: Evaluation and Control of the Optical Performance of Astronomical Telescopes"*, Proc. SPIE, **1236**, 575.

Espiard, J. and Favre, B.: 1970, *Nouv. Rev. d'Optique Appliquée*, **1**, 395.

Evans, D.C. and Breault, R.P.: 1984, *"Analytical Evaluation of the Diffuse Infrared Background Experiment for NASA's Cosmic Background Explorer"*, Proc. SPIE, **511**, 64.

Harned, N., Harned, R. and Melugin, R.: 1980, *"Alignment and Evaluation of the Cryogenic Corrected Infrared Astronomical Satellite (IRAS) Telescope"*, Proc. SPIE, **251**, 163.

Paquin, R.A. and Goggin, W.: 1972, *"Beryllium Mirror Technology"*, Perkin Elmer Report IS11393.

Thronson Jnr., H.A., Hawarden, T.G., Mountain, C.M., Davies, J.K., Lee, T.J. and Longair, M.: 1990, *Observatories in Earth Orbit and Beyond*, Proc. IAU Colloquium **123**, NASA/ESA.

Yoder Jnr., P.R.: 1986, *Opto-mechanical Systems Design*, Marcel Dekker Inc., New York.

Young, R.P.: 1976, *Opt. Eng*, **15**, 516.

CLOSED CYCLE COOLERS FOR SPACE APPLICATIONS

A. H. ORLOWSKA and T.W. BRADSHAW
Rutherford Appleton Laboratory, Chilton, Didcot, Oxon, OX11 OQX, U.K.

Abstract. Single stage Stirling coolers providing refrigeration at around 80 K have been developed for space use and are now being produced commercially. Development work is now concentrating on multistage coolers for temperatures below 30 K. This paper describes results from a two stage cooler built at the Rutherford Appleton Laboratory and preliminary tests on a closed cycle 4 K cooler.

1. Introduction

The performance of many detectors used in remote sensing is enhanced by reducing detector temperature to achieve a good signal to noise ratio. Cooling may also be required for other parts of space instruments: in the past this cooling has been accomplished by the use of solid or liquid cryogens but such systems have limited life and large mass. Closed cycle coolers can be used either to cool such instruments directly, or to increase the life of the cryogen by shield cooling of the cryostats.

Two instruments, the Along Track Scanning Radiometer (ATSR) on the European ERS-1 and the Improved Stratospheric and Mesospheric Sounder (ISAMS) on UARS were launched in 1991; both include single stage Stirling cycle refrigerators which cool their detectors to around 80 K. These coolers, which were designed for long life, were developed by Oxford University and the Rutherford Appleton Laboratory (RAL) (Delderfield *et al.*, 1985; Werrett *et al.*, 1986).

There is a demand for cooling at temperatures below 80 K. Many detectors (e.g. SIS) require cooling at around 4 K or below; such instruments presently use stored cryogens where the life is limited by the mass of cryogen and the heat leak into the dewar. These systems could have their lifetimes extended by the use of active shield cooling. A cooler producing a few hundred milliwatts at a temperature between 20 and 40 K might be suitable for such an application. For some instruments stored cryogens could be totally replaced by closed cycle coolers if a reliable long life system capable of producing a few milliwatts of cooling at about 4 K can be developed. Closed cycle coolers are expected to ease considerably the ground testing and prelaunch requirements for spacecraft since instruments can be launched warm and the coolers turned on once the spacecraft has reached orbit although some power during launch may be required to prevent cooler damage due to vibration. Optical systems would be designed to allow in-orbit cooldown and the resulting changes in alignment. Considerable savings in launch mass and hence cost over large cryostats can be envisaged although power requirements and heat dissipation must be considered.

Such developments would enable passively cooled spacecraft to achieve very long lifetimes and could cool selected components of the instruments to temper-

Space Science Reviews **61:** 233–240, 1992.

atures not attainable by passive cooling alone. Small cooling powers should be sufficient in such applications since heat leaks from the rest of the passively cooled system should be low.

2. Single Stage Coolers

Any mechanical cooling system designed to wholly or partially raplace a cryogen for space use must be extremely reliable, have long life, low vibration levels, and have modest power consumption. Size and weight are generally not critical when considering such systems in comparison to cryogens, but the cooler must be sufficiently robust to withstand launch conditions.

The Stirling cycle cooler has two mechanisms, a compressor and a displacer; both are driven sinusoidally by linear motors (with permanent magnets and moving coils) and employ clearance seals thereby avoiding friction between rubbing parts. The clearance seals are maintained by a diaphragm spring suspension system similar to that of the pressure modulators. Gas containment is achieved using gold O-ring seals. Power consumption is minimised by running at operating frequencies close to the compressor mechanical resonance frequency (determined mainly by gas dynamics within the cooler).

This cooler design evolved from that of the pressure modulators successfully flown on Nimbus 6, Nimbus 7, Nimbus F and Pioneer Venus (Curtis *et al.*, 1974; Taylor *et al.*, 1979). These mechanisms employed a piston driven by a linear motor and suspended on diaphragm springs. The absence of rubbing parts led to long life; the pressure modulator on Nimbus 6 ran for over 2×10^9 cycles (at approximately 20 Hz) before it was switched off.

The design of single stage coolers similar to that developed for the ISAMS instrument has been licensed to industry and the coolers are now available commercially. Comprehensive testing to pre-qualification levels has been negotiated successfully and several units are now on life test (one has run for over 36000 hours as of March 1992) (Scull and Jewell, 1991).

3. Two Stage Coolers

Refrigeration at low temperatures is power consuming for two reasons;

1. the Second Law of Thermodynamics limits the coefficient of performance (COP) of the system
2. many losses in the system are strong functions of temperature gradient and hence increase as the cold end temperature is lowered

These factors limit the performance of single stage coolers which give little useful refrigeration below about 40 K (Orlowska and Davey, 1987).

If the system can produce refrigeration at an intermediate temperature expansion space this will effectively lower the losses in the coldest stage since a part of the losses (such as regenerator losses and conduction) will have been intercepted at the

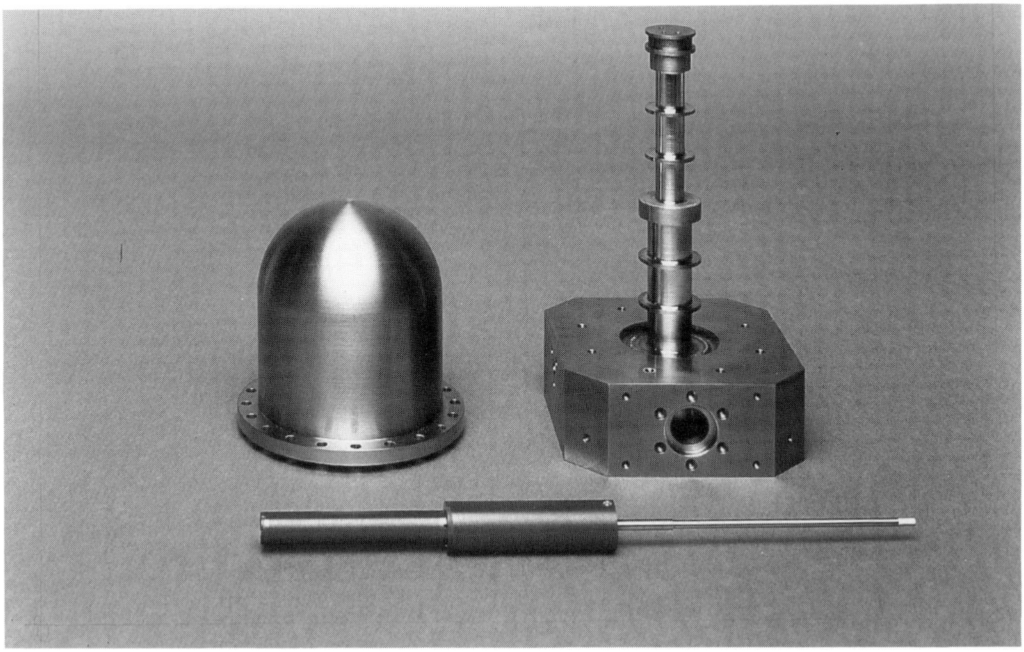

Fig. 1. The displacer and the cold finger of a two stage Stirling cycle cooler.

intermediate expansion space. This technique has been widely used in both Stirling and Gifford-McMahon cycles for ground based applications.

RAL has developed a two stage Stirling cycle cooler for space use. This cooler incorporates all the long life features of the single stage coolers, such as the clearance seals maintained by a diaphragm spring suspension system, but has a stepped displacer running in a stepped cold finger giving an annular expansion space at an intermediate temperature. The displacer and the cold finger in which it runs are shown in Figure 1. The two stage displacer has a large dead volume compared with that in single stage displacers, therefore two compressors are used to provide the alternating pressure. The compressors, mounted 'head to head' for momentum balance, are linked to the displacer by a short pipe. The complete cooler is shown in Figure 2.

Results of performance tests are shown in Figure 3 where the cooling power at a fill pressure of 0.9 MPa is plotted against temperature at different operating frequencies. The cooler has reached a base temperature of 16.2 K and provides about 60 mW of useful refrigeration at 20 K, and 320 mW at 30 K with a total input power to the compressors of 85 W. During these tests heat was rejected at 330 K; better performance would be expected for lower compressor heat sink temperatures. Further improvements have been made to the linear motor design

Fig. 2. A complete two stage Stirling cycle cooler.

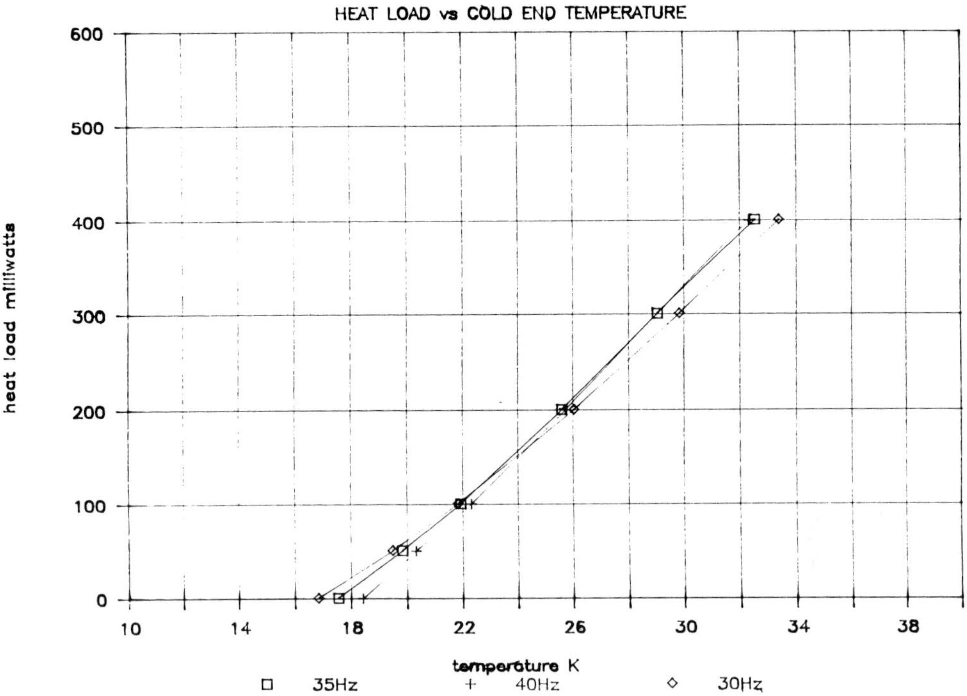

Fig. 3. Heat load versus cold end temperature.

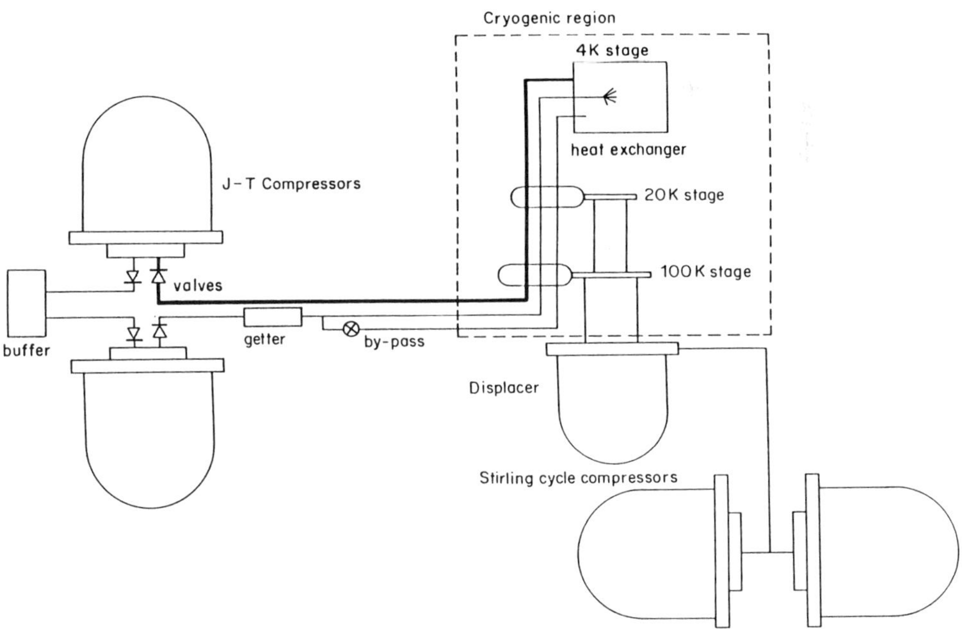

Fig. 4. Schematic of the 4 K cooler being developed at RAL.

Fig. 5. The complete 4 K cooler being developed at RAL.

which have led to a decrease in required input power (of approximately 15-20%).

Transfer of the two stage cooler technology to industry (British Aerospace plc) is currently taking place under a contract placed by the European Space Agency (ESA Contract number 9004/90/NL/FG placed with British Aerospace). This contract provides for environmental testing to pre-qualification levels including vibration and thermal vacuum tests which have been completed successfully. The two stage cooler is now starting a five year life test (Jones and Scull, 1991).

4. Cooling below 10K

All regenerative cycles including the Stirling cycle become inefficient at very low temperatures due mainly to regenerator ineffectiveness. As the temperature falls the heat capacity of most materials suitable for the manufacture of regenerator matrices also falls while the volumetric heat capacity of the helium flowing through the matrix rises.

There are materials with specific heat anomalies at low temperatures (e.g. certain Gadolinium rich alloys (Buschow *et al.*, 1975) and other rare earth compounds). Recently these have been used successfully in the regenerators of large, slow, ground based machines (Li *et al.*, 1990), but are only generally available as a powder or as spheres. For coolers such as those described here, which operate at relatively high frequencies (30-50 Hz), the pressure drop characteristics of the regenerator are particularly important and a more porous geometry is preferred. Work has begun at RAL to investigate such materials and their application to the long life space coolers.

In order to reach low temperatures (around 4 K) an additional non-regenerative cooling stage is generally required. RAL, under contract to the European Space Agency, is developing a Joule-Thomson (J-T) expansion stage using helium as the working fluid, precooled to about 25 K by the two stage Stirling cycle cooler. The high pressure gas is produced by two compressors identical to those used in the Stirling cycle but with the addition of reed valves to give a dc flow. This pair of compressors is also mounted with their axes in line in order to provide some momentum compensation (although in this case the piston diameters are not identical and the compressors encounter different gas dynamics so that the degree of momentum compensation is not as great as for the Stirling cycle compressors). Two such compressors with valves mounted in series give pressure ratios of up to 11 with a flow rate of 3 mg/s.

The high pressure gas passes through the inner tube of a series of concentric tube heat exchangers where it is cooled by the low pressure return gas from the J-T orifice. There are additional heat exchangers mounted on the expansion stages of the two stage cooler. A bypass of the coldest heat exchanger and expansion orifice is included for rapid cooldown of the J-T stage (patent application number 9004427). This is controlled by a room temperature valve which is closed once the J-T stage is well below the inversion temperature of helium. A schematic of the system is shown in Figure 4 and the complete 4 K cooler is shown in Figure 5. It should be noted that the precooler compressors must be mounted close to the displacer to achieve good performance, but that the JT compressors may be mounted a considerable distance from the cooler.

First results from the 4 K cooler are extremely promising. At present the lowest stable temperature that has been achieved is 4.05 K and the system provides 1.5 mW at about 4.2 K. Under these conditions the precooler input power was less than 70 W and the JT compressors were consuming 59 W (total input power $<$ 130 W). Work is continuing in order to improve this performance. The 4 K system technology is also being transferred to industry (Bae) under ESA contract (contract number 9004/90/NL/FG placed with British Aerospace). The final system will be completely sealed and will incorporate a hot reactive getter in the JT circuit to maintain gas purity.

5. Application of Cooler Technology to Passively Cooled Spacecraft

Both the two stage cooler and the 4 K machine could be used with advantage to lower the temperature of some elements of an otherwise passively cooled spacecraft.

From a thermodynamic viewpoint it would be very advantageous to run coolers at low ambient temperatures such as those achievable on the surface of such a satellite, but mechanical parameters such as resonant frequency and thermal contraction must be considered. In addition the problems of heat rejection, and the effect of coolers on the thermal environment of such a spacecraft have to be addressed.

Acknowledgements

This work has been supported by the European Space Agency (contract number 7198/87/NL/MA(SC)) and by the Science and Engineering Research Council. The authors wish to thank R. Wolfenden, R. Carter and W. Blakesley for their help in design, electronics and assembly.

References

Buschow, K.H.J. *et al.*: 1975, *Cryogenics* **Vol. 15**, 261-264

Curtis, P. *et al.*: 1974, *Proc. Roy. Soc. London* **A337**, 135

Delderfield, J. *et al.*: 1985, *Proc. SPIE* **589**

Jones, B.G. and Scull, S.R.: 1991, *Development and pre-qualification testing of a range of cryogenic coolers for applications from 4 K to 80 K. Proc. Cryocooler seminar I.Mech.Eng. London* ,

Li, R. *et al.*: 1990, *Adv.Cryo.Eng.* , 1183-11902

Orlowska, A. H. and Davey, G.: 1987, *Cryogenics* **27**, 645

Scull, S.R. and Jewell, C.: 1991, *Pre-qualification level testing of an 80 K Stirling cycle cooler. ESTEC environmental conference, Florence* **ESA-SP 324**,

Taylor, F. W. *et al.*: 1979, *Appl. Opt.* **18**, 3893

Werrett, S. T. *et al.*: 1986, *Adv. Cryo. Eng.* **31**, 791

ASPECTS OF DATA FLOW FOR AN EDISON-TYPE OBSERVATORY

ALAN PENNY
Rutherford Appleton Laboratory, Chilton, Didcot, Oxon, OX11 0QX, U.K.

1. Introduction

This paper discusses the design of the flow of data for an 'Edison'-type space mission. The concept is one of a large (2–3m) diameter telescope for use in the near (2–50 μm) infrared and having a mission lifetime of ten to twenty years. This would be achieved using a passively cooled mirror and an instrumental package whose passive cooling is aided by a mechanical cooler. The data-flow to and from such a telescope would be large and thus the data-flow design merits careful consideration.

There is a tendency in the very early days of the design of a new space mission to concentrate on the spacecraft, telescope, and instruments, and to leave the overall system design till later. However the aim of a mission is not to have innovative components, but to do new astronomy, and it is with this in mind that the system design should be done.

A system design will show that in addition to these three elements, the process of data-flow is an important part of the system. Blockages, limitations and delays in the process of data movement will affect the astronomy done with the observatory, and whilst it is difficult, expensive, and time-consuming to design the data-flow in from the start, it is even more difficult, expensive, and time-consuming to correct any deficiencies afterwards. It can even be impossible to do so.

Data-flow can be considered as consisting of four parts, as shown in Table I.

Each component of this flow must be matched to the astronomical aims of the mission.

This paper discusses some of the lessons that may be learnt from past missions and some possible new developments.

2. Lessons from the Past

Because of its anticipated long life-time, an Edison-type mission is one that operates like an observatory rather than like a PI project. To understand the implications of this, similar missions done in the past may be studied. The main relevant missions are two space missions, IUE and HST, and the 'missions' of the ground-based observatories. The all-sky survey from IRAS also has some relevance, as has the planned SOHO mission. Some aspects of these missions give a useful guide to future practice.

Space Science Reviews **61:** 241–246, 1992.

TABLE I

Direction	Purpose	Path	
		From	To
Inwards	Get information and instructions from community	Community	Science Institute
Upwards	Control spacecraft and instruments	Science Institute Ground Station Spacecraft On-board Computer	Ground Station Spacecraft On-board Computer Instruments
Downwards	Take, reduce, analyse and store engineering and astronomical data	Instruments On-board Computer Spacecraft Ground Station	On-board Computer Spacecraft Ground Station Science Institute
Outwards	Supply data and analysis tools to community	Science Institute	Community

2.1. IUE

IUE has been a very successful mission, not only because of the long life-time of the spacecraft, but also because the data flow has worked well. Some particular aspects of the IUE mission are:-

- Community Participation: A good interaction with the community. There was and is much input into the design, mission definition, determination of scheduling priorities, design of reduction software and calibration techniques.

- Mission operation: The simple mode of operation has enabled efficient use of the spacecraft. Under direct astronomer control, the scheduling, communications and control has been efficient and flexible. Further, as the mission has progressed, the kind of astronomy done has varied, but it has been possible to adapt the observation methods and controls to allow for this.

- Calibration: During the course of the mission, the calibrations of the instruments have evolved, with both calibration observations and analysis software still being developed after ten years of mission. There was recognition that, initially, the calibration was limited, and greater efforts have been made in the later stages.

- Archive: The archive, with over a decade's worth of observations is proving very useful. The usage of the archive is such that more exposures are now extracted each year from the archive than are taken with the spacecraft.

2.2. HST

HST has only had a year in operation, and so the lessons to be learnt are only just emerging. However, because of its major data flow problems, over the last ten years there has been much work devoted to this problem, and some lessons have emerged clearly. Amongst these are:-

- Design: The need for all aspects of the data-flow to be astronomer-driven and -overseen, from the design stage onwards. Much expensive rewriting has been necessary as the project has evolved, and unplanned for facilities added, to some extent because of the initial isolation of the various groups working on the data flow. The need for flexibility in the software for upgrades has become paramount.
- Operations: The high cost of a complex schedule, both in effort and loss of capability, has become apparent. Also the importance of a science institute to control the data flow and to interface with the community has been demonstrated by the success of the STScI.
- Data-Flow: The limitations on data-flow imposed by the communication network restricts the full parallel usage of the instruments, and also imposes significant overheads on scheduling. They also limit the use of new developments in observing techniques, and the types of instruments that can be installed in the future.
- Calibration: The need for extensive calibration of the telescope and instruments has become apparent. This has involved both the development of new analysis software, and the development of new calibration techniques and exposures.
- Archive: The archive has, after the first year of operation, some 25,000 public domain exposures. This has demonstrated that the deployment of a powerful archiving system was justified, and implies that the archive will get substantial use.

2.3. Ground-based Observatories

From ground-based observatories comes much modern experience of data flow. Amongst the lessons are:-

- Integration: The instruments have to be designed to allow for both the data flow and the software that will be used to control them. Also the integration of telescopes and instruments with the control computers into a coherent whole has become essential.
- Calibration: For ground-based observations the fact that rarely are there standard calibrations has led both to limited accuracy in the calibrations and little

re-use of observations from the archive. While to some extent this is due to ground-specific problems in calibration, the low priority given to calibrations in the scheduling has exacerbated the problem.

Archive: In addition to the poor calibration of data in observatory archives, the archives themselves tend to be in different formats, meaning that their use is difficult, indicating that a standard archive system would be desirable.

2.4. IRAS

From the IRAS survey, the most relevant lesson is the need for extensive calibrations and advanced data reduction tools. Some ten years after the mission there are still the 'stripes' in the data caused by fluctuations which have not been perfectly calibrated out, and the use of newer reduction tools, such as Maximum Entropy methods, are still being developed. The large IPAC effort on post-launch processing shows how much effort is needed to analyse the data from this type of mission.

2.5. SOHO

SOHO is a solar observatory which will be stationed at the Lagrangian point between the Earth and the Sun. It is planned to have a data rate of 5kbytes/sec, with a peak short-term rate of 20kbytes/sec. Even this rate depends on use of the NASA deep-space network, and this is only available for a proportion of the time. For an Edison-type observatory, at a similar distance, the much higher data rates envisaged will imply that a substantially more powerful data system will be needed.

3. Why the Future will be Different from the Past

3.1. General

A major change for future infra-red missions will be that the data sets and data rates will be much larger than for past missions. Hence the data-handling system should be designed so as to accommodate the maximum data set size and data rate from any instrument that might be installed on the spacecraft. An example of a failure to meet this criterion is the HST communication system, designed in the early eighties, which cannot handle the full data-flow from the near infra-red instrument (NICMOS), presently being built for installation later this decade.

Fortunately, in an Edison-type mission the instruments will not be replaced during the mission, so the maximum data-rate from the instruments should be known in advance. However, there should still be ample on-board storage for large data sets, both for temporary storage, and for on-board processing, and sufficient on-board processing power to both handle the data and do any pre-processing that may be desirable. There should in fact be a large over-estimate of the storage and power needed, as experience shows that during the mission new techniques of observing and telescope control will be developed which will make new demands on storage and power. It is important to realise that in any case these needs will be larger than in previous missions, not only because the infra-red instruments have inherently high needs, but also because in general instrument and telescope data needs are increasing with time for all missions.

On the ground, new developments in data-handling will include higher speed communications, more powerful calibration and reduction software, more widespread use of standard archive and data formats and of standard software tools. This will permit a blurring of the distinction between different locations in a network, which will in turn give both advantages (a faster flow of data and reduction techniques between groups) and disadvantages (tighter control needed to ensure uniformity in results). More powerful archives, capable of storing terabytes of data with fast recall, giving greater and faster access, will lead to greater use and a demand for good calibrations.

Overall, the fraction of total mission costs devoted to the data-flow in all its aspects will increase.

3.2. Calibration

For a long life-time mission, calibration will need to be an ongoing process. There should ideally be a greater fraction of mission time devoted to calibration observations, and there should be different methods available (such as on-board sources) to cross check the methods. As the mission continues these calibration methods will evolve, and this will permit the recalibration of earlier observations in the archive. Also, the use of new methods of analysing the data, such a Maximum Entropy methods, will continue to be developed.

3.3. A Science Institute

The experience of the STScI shows the need for some institution or location to coordinate (or perhaps undertake) the various stages of the data-flow. The need for such coordination of the various efforts is high, because the high-speed networks will blur the boundaries between different groups in the community and the boundary between those groups and the mission. This coordination will affect all parts of the planning and control of the mission, enabling the scientific instruments, data formats, reduction tools, archives, etc., to be developed and used as a coherent whole.

3.4. Spacecraft and Communications

The high data rate envisaged for this type of mission will lead to new demands on the communication system. If the spacecraft is in a distant orbit, at a Lagrangian point, then a combination of more powerful antenna on the spacecraft, more sophisticated data compression and coding methods, and suitable ground antennae will be needed. Also the communication between the ground antennae (which for a Lagrangian-point orbit will need to be spaced round the Earth) and the science institute, both for sending commands and for the reception of data, will need to be adequate.

A listing of the requirements for existing and planned missions can be compared

TABLE II

Component	Mission				
	IRAS	HST	ISO	SOHO	EDISON
Commanding Rate (Kbit/s)	1		2	2	(2)
Data Rate – Science (Max) (Mbit/s)	1	1	0.024	0.16	(10)
Data Rate – Engineering (Max) (Mbit/s)	0.004		0.012	0.004	(0.01)
Availability of Maximum Rates (% of time)	1	20	60	100	(100)
On-board Computer Memory (Mbyte)	0.068		0.24	4	(100)
On-board Storage (Mbyte)	50		0	1000	(10000)

to very rough estimates for EDISON (see Table II).

The very rough EDISON estimates are open to discussion. They are based on the assumption the main difference between it and similar missions (ISO and SOHO) is that EDISON will have a 256×256 detector, needing fast read-out rates. This will mean that the on-board processing and storage capacities and the data-rate down from the satellite will have to be significantly higher than for previous missions. The need for fast read-out and the preservation of as much of the original data as possible mean that the data rates will have to be significantly higher than even for the HST mission. The large on-board storage will not be essential if 100 percent coverage can be achieved, but should be available as a backup for the case that such complete coverage is not maintained.

There are implications on the spacecraft design because of this problem. A large antenna will have thermal effects and weight implications. The large power consumption of a powerful transmitter will have other implications, as will a powerful computer and data storage system.

EDISON DISCUSSION

MODERATOR: MALCOLM LONGAIR

LONGAIR – The word moderator can mean many different things. There is the Moderator of the Kirk of Scotland who is first among equals. There is the moderator in nuclear reactors which slows down the neutrons to encourage them to produce a controlled chain-reaction!

Many of us have been thinking to a greater or lesser extent about Edison and Poirot type missions. Harley Thronson's presentation was extremely helpful in setting the political atmosphere in which one can attempt to do something. But before we get on to this I thought that we should discuss the technical points in more detail.

We have this whole question of scientific objectives and I regard what we heard as basically scraping the surface of the scientific case. It is absolutely crucial to plan the detailed scientific case and this must be one of the major priorities. You really want some particular thing that Edison would do that could not possibly be done by anything else.

THRONSON – Sure, something *good* that can't be done by anything else.

LONGAIR – I have this theorem for space astronomy which is that the scientific cases for all large projects are the same. This is because every large project is going to work out Hubble's constant, or everyone is going to work out the deceleration parameter or everyone is going to find protogalaxies.

EMERY – I wonder whether the discipline has matured to the point now where we could actually focus down a little bit and specifically address the questions of very high spatial resolution for this particular mission. One of the dangers at this stage is that you get invaded with requests from all sorts of astronomical interests and then you start trying to be all things to all people. The community itself has to recognise that there is a very good case for focusing in on a few aspects of performance which match very high priority items of the scientific case.

THRONSON – Let me repeat the philosophies we first came up with. The philosophy generally was that no matter what your science case was the bigger the aperture of the telescope the better since astronomy is a photon-limited science. We were saying yesterday that we chose $2 - 2\frac{1}{2}$m as being the largest mirror you can get away with which can fit into an Ariane or an Atlas and maybe $3\frac{1}{2}$m could be

Space Science Reviews **61:** 247–264, 1992.

shoe-horned into an Ariane 5 or a Titan. We felt that anything bigger would probably be very difficult and couldn't really be done in the timescale we are talking about.

THOMPSON – First of all I would like to say that I usually look at meetings like this as an opportunity for free and open discussion. I shall also say, since it is on the record, that I think this is a fabulous idea and I am glad people are talking about it and I do believe in radiative cooling. I also believe in radiative heating and I invite Harley to take a walk with me in the desert in Tucson so that he will be able to feel the effects of radiative heating. Also I think that radiative cooling has a tremendous advantage in the wavelength region going up to 5 microns in that the detectors will run to the temperatures, which I actually believe you can get to, of about 60K. For this kind of space mission the dominant background is emission from zodiacal dust and by 10 μm the zodiacal background is coming up. The region from 1 to 5 μm is where you get the very deepest view of the universe and that is scientifically very important. So my proposal would be for a large telescope that did not need any more cooling and which dedicated itself to cosmological and star formation projects; that is if you want to match this really neat idea of radiative cooling with a mission which is only about a factor of 2 more complex than HST. You should realise that you are in a big mission here, even if the mirror is only $2\frac{1}{2}$ metres. In this particular part of the spectrum I think you could get a very, very important scientific justification and do science that no-one else could do. There is no project that has this sensitivity because of its size, its angular resolution and for exploiting the unique idea of radiative cooling.

LONGAIR – Rodger, would you like to describe for us the views in the US discussions of the large Next Generation Space Telescope and where the infrared would fit in. I think that this would be helpful.

THOMPSON – OK. There has been an effort to look at the Next Generation Space Telescope and we saw that on Harley's graphs. You might think that given our experience on the first one, it takes a lot of guts to go for the second one, but remember that it takes 20 or 30 years to get these missions going. It has become clear that no-one in that group wanted to go below 8 metres. First of all, 8 metres was becoming a small telescope, and some people were starting to lose interest in 8 metres. The second thing that became very clear is that the needs of the optical/uv versus the infrared were becoming contradictory and it may be impossible to have a telescope that serves all people. First the infrared people want the telescope as cold as possible. The low contamination levels that you have to have, especially for uv telescopes, may put very stringent constraints on an infrared telescope. The coatings we would like of course are golden coated telescopes, and the uv/optical people are not really keen on that sort of thing. Another issue is the temperature stability of the mirror and making sure that you know what temperature you are going to be at before you get there so that you can figure the mirror properly for that particular temperature. You do everything you can to achieve as low a temperature as possible and then you add a small amount of heat to the telescope. This will drive it up somewhat above the temperature it would go to normally but enable you to have control over the temperature of the telescope. All the details could be put in and 60K might not be a bad temperature to aim for.

THRONSON – This is a comment about the science. We have considered the arguments by other people about the importance of the 3 to 4 micron region for cosmology. Here is one of the base-line Edison science figures (Figure 4 in Thronson *et al.*; this volume). This is not cosmology but this will give you an example of the importance of the science in the very few micron regime and show the sensitivity of Edison at $2\frac{1}{2}$ metres for a 40K or 50K system. The graph shows that, with broad band photometry, you could detect Type I supernovae at a redshift $z = 4$, if such things exist. In arguing the scientific case you've got to be really careful about the near infrared because, although you might get a lot of information about cosmological questions, it may not be the information you want 10 years from now. Two other communities that we considered were the stellar community and the solar system community. The stellar people, for example, pointed out the importance of around 10 microns for brown dwarfs. We don't propose this mission as a brown dwarf detector, as there is a very good chance that very many brown dwarfs will have been detected a decade from now. Instead we propose that spectrometers on Edison would observe brown dwarfs. Brown dwarfs possess a wide range of diagnostic atmosphere features that exist through the middle infrared. Notice the large number of communities that we would serve with Edison:– the star formation community, the stellar community, the solar system community, the cosmological and high redshift extragalactic community. Each of these types of science are as astronomically important as any other. This is why we argue in favour of a general broad wavelength mission as opposed to just an isolated, highly specific, mission. You would be wasting a lot of glass if we do not open up the wavelength range.

CHASE – What about even longer wavelengths? Tim was talking about the advantage in a large mission of reducing the confusion problem at longer wavelengths. At what point do you stop broadening it out? Is it sensible to go to 500 microns?

THRONSON – I think it certainly might but I think that the very good point, the summary point, that Rodger made about the uv and optical communities realising that there are too many conflicting technical demands, applies if you go to too long a wavelength in the infrared. Eventually for example, pointing constraints, aperture decisions, and so on, start making Edison much less attractive than something like FIRST. My guess is that several tens of microns would be the longest wavelength.

LONGAIR – Again I come back to the point that you are basically building a passively cooled HST and that means that we are talking about a very large project. Can I take a separate question and that is the possibility of a passively cooled IRAS. This is now talking about a very much less expensive mission which would be dedicated to doing all-sky surveys roughly a thousand times better than IRAS. Now, presumably one can apply all of the concepts of passive cooling to this mission and place it in the optimum orbit so it's pointing away from the Sun at all times. Is there some feeling that this is an important thing for people to think about? This would be a completely different order of magnitude project from the big one.

WALKER – There is a lot of good science you can do with another survey mission. That would be a good stand-alone mission which is not so expensive, using the new technology and just going further in wavelength.

THRONSON – I think a survey like that would be extremely exciting. Maybe the Japanese mission would be something like that.

HAYASHI – I think that at present we can only do a very small telescope for a dedicated survey such as a passively cooled IRAS. It would be complementary to your Edison pointed mission.

HAWARDEN – I have been thinking about such a mission. The constraints are to launch round about the year 2000, with overall maximum diameter 2m, and weight 600 kilograms and to be placed in a low Earth Sun-synchronous orbit. I am fairly confident in saying we could get a 1-m class telescope down to 40-45K in a suitable orbit. We should encourage our colleagues in Japan to take this opportunity to go and build something which could be both small and beautiful.

THOMPSON – I just wanted to make one comment on the advantages of parochialism in the sense that the 1 to 5 micron region covers a factor of 5 in wavelength which is the same wavelength coverage the HST had originally. It covers at its short and long wavelength ends very different types of science. The graphs that you have shown, Harley, show very clearly that the very largest telescopes have advantages over other systems within that region. Then follow this by the fact that you are so neatly tied to the technology of radiative cooling that it gets you where you want to go without any of the yet unproven cooler technologies. Certainly the infrared astronomers I know look at that as a very narrow region, but in reality it is a very broad region of wavelength and of science.

LONGAIR – Let's go on to talk about some technical issues. Let us grasp the key to this project – passive cooling. If the telescope is shielded from solar and terrestrial radiation will it cool to roughly the microwave background radiation temperature?

THOMPSON – Well almost, except it will cool to the radiation you receive from zodiacal dust.

LONGAIR – To order of magnitude it has got to be round about $T^4 \times 10^{-6}$ and T $\sim 60 - 100$K is the temperature of the zodiacal dust. If you put all your effort into shielding it would get you down to the order of 10K. If we accept these sort of ball park figures are going to be right, how important is this question about keeping temperature stability? How important is it to make sure that the fluctuations in temperature really are quite small?

THOMPSON – I think we have two responses. First of all temperature fluctuations are a function of a variety of spacecraft issues – turning instruments on, turning instruments off, orientation, loss of orientation between the Earth, the Sun and the telescope. The quick response is to remember that many of the observations which will be undertaken by the spacecraft will be of wavelengths much shorter than the black body peak, way down on the Wien side. If such temperature fluctuations exist, you have the option of observing at wavelengths much shorter than the black body peak.

LONGAIR – Now are we all happy about that, but suppose that you are on, or close to, the black body peak. To what extent does that affect the science in the region of about say 20-40 microns?

THOMPSON – Then you've got a real problem. Notice that the cooling time is very much different from the heating time. This is one argument in favour of fairly large amounts of mass so one would expect the temperature fluctuations to be very small fractions of a degree.

DAVIES – There is also the orbit. Ever since we started thinking about this we have been pushing it into steadily higher and higher orbits, and if we finish up in L2 the likelihood of large temperature fluctuations will get smaller, because, for example, we do not have perigee passages to worry about.

THOMPSON – Regarding the accuracy of the pointing, I would say you have to learn a lesson from HST – the temperature variations of those outside members, they expand and they contract and they twang and they move. You require a very, very accurate analysis of the mechanical coupling between those members. I think that's going to be your worst problem in this situation because you are deliberately putting things out there to shield you from radiation and absorb that power. That is a big argument against low earth orbit for this type of system, where you have all sorts of mechanical flexures as a result of frequent night and day.

LONGAIR – Does anyone want to talk about the passive cooling?

DAVIES – I have a question about technology and hybrid cooling. We are trying to design a mission here with a long lifetime. But something is going to limit the spacecraft's lifetime even if it is detector obsolescence or depletion of attitude control gas to unload the gyros. I wonder if we might use cryogens to cool the focal plane on the grounds that if it doesn't have to cool the telescope a small pot of cryogens could keep your instruments cold for years. There are cryogens in NICMOS, yet nobody thinks of NICMOS being a short duration instrument and there were solid cryogens on one of the HEAO satellites. I am not suggesting that it is a good idea or a bad idea, I just want to bring it into the discussion because it is an engineering issue which we might like to look at.

CHASE – I think it becomes a science issue as well in terms of the acceptable lifetime.

THOMPSON – Cryogenics are very good if you start off at the temperatures you are talking about, and any cryogenic engineer who is told that he has got 60-40K rather than the 270 or the 230K they usually have to start with is in seventh heaven. So that's how we have a long lifetime. I would suspect that a liquid helium type of system could last for a very long time, several years, if it was only cooling instruments or detectors inside. The other point, and I hate to bring up another negative, but the problem with mechanical coolers are they are all very inefficient and they have to dump heat some place. They are generally just the wrong thing to put in a radiatively cooled space telescope. Given the state of the coolers and the fundamental problems that they are inefficient and create heat and you have to dump it, that's going to be a really big problem.

DAVIES – The second thing is that this satellite will need a cover of some kind when launching. I mean to keep the dirt out. Can you put a cryogen tank in that to get you some pre-cooling and then jettison it?

HAWARDEN – Let me cover the pre-cooling idea. I think a rough calculation of how much cryogen is needed to cool the mirror down to 40K comes out to about 4000 litres of hydrogen – it does get to very big numbers.

LONGAIR – We'll come back to coolers again. Let us grasp the one more nettle of the pointing, which we are slightly assuming is no problem. Again, remember in my introductory remarks I made the point that whilst you've got a certain diameter with a certain wavelength then you get a certain pointing requirement.

THRONSON – I don't think that the requirements are as demanding as the HST.

EMERY – Can I make a couple of points. The first thing is that you've got to make a decision about whether the pointing involves some kind of adjustments of the optics in the telescope itself. If so, then that's got to be folded into the telescope design. Some of the decisions and trade-offs that were shown to us may well be substantially influenced by that. I think the other point is I haven't heard anybody make a conscious judgement about whether you need the diffraction limited performance at the shortest wavelength.

LONGAIR – I think there are at least two people in this room who will cancel this project if it doesn't image perfectly. I would almost make it a requirement of the project that you've got to have exquisite imaging and pointing to 2 microns.

WELLS – There is a question of absolute pointing as distinct from pointing drift. If the pointing drift is less than half a pixel within an integration time then it doesn't matter because you can measure the drift, you don't control it and then you just stop integrating your detectors when you've got to half a pixel off. So maybe this is a question to the spacecraft pointing side of it, can you control the drift rather than the absolute pointing. Does that make it easier?

LONGAIR – I think you are going to lose an awful lot if you are going to try to do that because you are dealing with diffraction limited images and this fuzziness simply makes your photometry that much more difficult. The fine guidance system on the HST does work but my feeling is that we should be able to improve on a system designed on the basis of the available technology in 1977. I believe that one can do it more simply nowadays with techniques using imaging detectors.

LONGAIR – One question which I would like to raise is the cost diameter relationship for telescopes of this nature. Rodger told me that the only correlation he was aware of among the professionals, between the cost of an instrument and what it actually looked like in the end, was a proportionality between the number of drawings and the cost of the project. Would you like to amplify on that Rodger, because if it is true it is a rather profound statement.

THOMPSON – You find the design costs are the main drivers of a project. It is not the materials or the actual manufacturing process that is the driver. It is the number of drawings and complexity that drives the cost. In that sense, a 4 metre telescope doesn't generally have many more drawings than a 2 metre telescope.

THRONSON – It seems to me that the contractor would benefit greatly by using very small drawings!

HUMPHRIES – Could I just say something about how these power laws arise. At least in ground based telescopes the overall power law index of 2.6 or 2.7 on diameter is the result of a lot of contributions, each with different power law dependencies. Whilst the number of drawings may not vary much with telescope size there are other things that go as mass or volume that obviously have a diameter cubed relationship. There are some things like the optical polishing that go as the square of the diameter and there are yet other things like the amount of electrical wiring in a telescope where the dependence will be somewhat more linear. You've got polynomial terms adding to give the overall power law relationship observed. What it is for a space telescope I don't know. I presume either NASA or ESO do, but generally if you want to reduce costs by breaking the mould from what's been done in the past, then you've got to find some way of reducing the contributions of the largest terms in the polynomial.

ERICKSON – I wanted to ask Rodger what happens to the cost versus mass relationship. Lockheed gave us an estimate for SIRTF 10 years ago or so. They showed this drawing of cost versus mass for hardware they had built and launched and it seemed to be a very good linear relationship.

THOMPSON – There also was a relationship in the sense that if you plotted the number of drawings per mass then the more massive space projects tend to be more complex. It wasn't just the mass that was the problem, it was the design costs. I think the difference between ground based telescopes and space telescopes is first that the numerical factor of cost that you put in front of the design when you deal with an aerospace firm is much higher, and the quality control and all the things you have to do for each piece puts a much bigger factor in front than you do costing a University or Observatory doing a ground based telescope. Also a major factor in ground based telescopes is the building you put it in, and buildings tend to go as the cube. We found that the easiest way to reduce costs for our ground based telescopes was to make fast telescopes which required smaller buildings. That changes the power factor quite a bit. So it is just where the costs are in a space project that are different from what we are used to.

LONGAIR – I've got closed cycle coolers as the next thing on my list for discussion as clearly being a super thing. We do sums as classical thermodynamicists to work out how much power you require to drive a cooler. Can we trust these calculations when we do them for closed cycle coolers or have we got to put in an extra fudge factor?

ORLOWSKA – On that you need extra fudge factors. For instance the 80K coolers probably give you 9-10% Carnot efficiency, with 20K coolers you are down at 3-

5% Carnot efficiency and with 16K coolers you are even more inefficient, close to 0%. There is also the size factor. Generally speaking again, the larger the better. There are other small space coolers which, although they do well for their size, in absolute terms have low efficiency.

THOMPSON – For a radiatively cooled telescope you do not require a very fast cooling technique. The difference between cryogenics and closed cycle coolers is that for the cryogenics you have dumped your heat before you even start the mission. For closed cycle coolers you have an efficiency factor. You are using much more energy than you are removing from the bath and you have to dump that energy. You have to add that energy to the radiative cooling requirements and so when you are talking about a radiatively cooled telescope I think you have to be very careful about talking about closed cycle coolers because I don't think the match is correct. The fact that you do have a radiatively cooled telescope makes cryogenics last a lot longer. This is because it is the temperature gradient between what the cryogens are at and what the outside is at that matters. I think you get locked in the wrong way by saying you have to use closed cycle coolers.

CHASE – There is possible advantage in hybrid coolers. If you have cryogens to cool the focal plane instrumentation you can in principle use boil off gas to stabilise the temperature of the primary.

ORLOWSKA – But if you have a small cryogen tank that you want to last for years your boil off has got to be very small.

MORGOVWNIK – Can you have the hot parts of the coolers in the service module?

ORLOWSKA – I think that's certainly true of the 4K coolers because you can have that very distant from the object that you are trying to cool. The other very good thing about it is that the coolers for the long wave detectors can be switched off if you've got a vibration problem with your short wave detectors. Then when you want your long wave detectors you turn your cooler on again. That can easily be done.

LONGAIR – All the long wavelength detectors I know need to be cooled down to 4K or 8K. There is clearly a correlation between the longest wavelength at which we wish to operate and the cooling requirements.

THRONSON – The desire for long wavelength performance leads to a requirement for a refrigerator.

RUSSELL – It is important to establish the operating range of wavelengths. The long wave cut off should be somewhere between 40 microns and say 500 microns. The exact wavelength is important in regard to overlaps with other missions.

HAWARDEN – I would like to summarise what I know because obviously this is a very crucial area. Rodger has been talking in terms of indium antimonide working out to 5 microns at 60K. I am trying to sell 40K. I am a little dubious about 60K simply because of our experience with UKIRT instrumentation where we try to stabilise at 35K for optimum performance. Now there is a regime between 5 μm and about 32 μm, where silicon runs out. Let's assume that we can get to 32 μm with doped silicon. Providentially that happens to be roughly where the radiatively cooled 40K 2.5m telescope gets overtaken by the cryogenically cooled 1m telescope.

ERICKSON – I think that silicon BIBs that we have run out to 27 μm and that they operate best at about 8K. We also operate beryllium doped germanium detectors between 25 and 50 microns and they like to run about 3.5K. Gallium doped germanium runs from say 40 microns up to 120 microns and works at around 3K.

LONGAIR – Our closed cycle cooler friends, you hear what we need!

ORLOWSKA – We have a programme to get to 2.5K. Below that it is difficult, you need a 4th stage. It's not so much the size, it's the fact that our 4K cooler has a limited pre-cooling power and really in order to build another stage on it you need to build a slightly bigger 4K cooler.

LONGAIR – The next item on my list is instrumentation. Now we've talked very little about that, I think that is probably appropriate at this stage because our wish list is to have massive arrays, massive two-dimensional spectrographs and so forth. Do you think we need any more input at this stage? I personally don't think we do. Good. Next thing I have written down is orbits and I hadn't appreciated until this morning the miracles that the L2 orbit provides.

HAWARDEN – The L2 orbit is ideal. It's on the right side of the Earth so that both the Earth and the Sun are behind the sunshade. You can then paint the back of the spacecraft black, with the consequence that it's going to get even colder. However, we may not be able to launch more than about 3200 kilograms into L2. At the moment our best guess for our strawman Edison is about 4200 kilograms so this presents a problem. Tradeoffs have to be made and I would be prepared to go down to 2 metres at 20K to meet the weight limit. I think L2 is the natural home of this satellite.

THRONSON – Do the L2 orbits present a hazard in terms of trapped debris?

FARROW – No. The orbits are not fully stable and so you need to do some station-keeping.

FURNISS – What about data rates? Do you need data relay stations since you are not in geosynchronous orbit?

FARROW – These are problems for any high earth orbit mission and are soluble for an international project. The L2 orbit has been studied by SIRTF. What was the result?

THRONSON – The project went to JPL! (laughter)

THOMPSON – You are going to have to worry about removal of radiation events and the long term stability of detectors, so it's something you have to take into account.

ERICKSON – Silicon BIB detectors are much less radiation sensitive than the bulk detectors.

LONGAIR – I'm almost through my list of points for discussion. Optical design is the next subject on my list. I got a strong impression looking at that list that there weren't any overwhelming reasons to go for off-axis design.

ATAD – On-axis designs are best in terms of manufacturing.

HAWARDEN – Other parties have criticised this idea on the grounds that you can't make a 2.5m high accuracy mirror and cool it down to 40K. Since it is obviously possible to make a 2.5 high accuracy dish without cooling it down to 40K that issue must be a matter of the non-uniformity of the material and its cryogenic performance. This is clearly an area where further work needs to be done.

ERICKSON – I am sceptical. I really think that part of your technology demonstrations requires some wealthy party to build a full scale test facility for this. The HST was heavily criticised for not having done an end-to-end test because it was deemed too expensive. At this point we all see the error of our ways and I think that we will have a hard time persuading people that we shouldn't do full scale tests.

HAWARDEN – I don't see how you can do zero gravity flotation test and a cryogenic test but you are going to need to do both.

LONGAIR – Do you want to raise this whole question of metal mirrors?

ATAD – Metal mirrors are attractive since everything expands or contracts uniformly.

THOMPSON – We have experience of using 2 metal mirrors. I think it only works up to about a metre but they are made of a very interesting material and they can be tuned to the same coefficients of expansion as carbon fibre for example. Other composite materials can be tuned to the coefficient of expansion that you want. I think all caution should be put on figuring the mirror, testing the mirror and, once again, this is where you have to get the temperature that you are going to reach correct. I hate to say it but I think you are going to have to use up a few of your degrees and temperature control the mirror at just above the minimum temperature.

ERICKSON – I want to ask Tim if he did a calculation for the cooling with a glass mirror as well as with an aluminium mirror.

HAWARDEN – I have calculated cooling times for three models referring to different assemblies that one might consider in a telescope tube, involving a meniscus mirror of some composition, a support system for the mirror primarily to ensure launch survival and a mass of aluminium to represent the instruments. The aluminium of the telescope tube is not structural, it is just a conductor with a radiative top end. What happens in the initial stages after you open the lid is that the telescope tube more or less freefalls down in temperature. Its cooling rate is controlled only by its own thermal inertia. The aluminium in my models is 2mm thick and it cools rather quickly. In 20 hours you have got a 130K temperature difference between the top end and the mirror. This is an interesting situation to be in with a delicate optical system, particularly the secondary mirror attached to the top end of the tube. This is why I like the idea of tuning the expansion coefficient to that of carbon fibres. After a while the cooling process is governed by the conductivity of the tube. At this stage the heat being radiated as T^4 from the tube is in a slowly changing state of equilibrium with the heat being conducted down the tube and the cooling rate depends on the conductivity and the thermal capacity of the mirror, etc. Finally the temperature difference down the tube becomes quite small and everything is governed by T^4 cooling.

The most realistic model I have produced so far has got a lot of carbon fibre support structure behind a quartz mirror. With this sort of support structure you are going to be touching the mirror in a lot of places with something like carbon fibre. One could put in conductive straps between the mirror and the tube. I don't think it is difficult to arrange good thermal connections between the mirror and the tube. So the main criteria for fast cooling are going to be the conductivity of the telescope tube, the size of the aperture and the thermal inertia of the mass at the back. If you really want to stick your neck out then choose a beryllium mirror which could be of lower mass as the material is very stiff. The thermal capacity of this mirror drops so low that it gets to 70K in about 40 days. On the other hand beryllium mirrors may have other drawbacks like they are probably not manufacturable.

LONGAIR – There's only one last item I have got on my list here, unless anyone has got anything that they desperately wish to raise, and that is something called the MANIFESTO. Harley, why don't you talk on the Manifesto and I also invite Harley to make any concluding remarks.

THRONSON – My concluding remarks are to thank you very much for coming. We appreciate the advice, support and criticism and hope that it will continue. As to the Manifesto – there is a real problem here as infrared and submillimetre space astronomy has been very, very difficult to get going not only for technical and budgetary reasons but also for political and administrative reasons. As many of you know the Bahcall committee in the US, and I think there is a parallel view on the other side of the Atlantic, refers to the 1990s as the decade of infrared astronomy. The international and national space missions planned for the 1990s show six infrared and submillimetre space missions. This is a good range of missions if they go in something like the timescale envisaged – it is a very exciting time for infrared and submillimetre astronomy. However, note that beyond about 1995 there are no further missions under development or serious study in the infrared and submillimetre. I want to contrast that with the listing of missions from a recent American Astronomical Society Newsletter which showed the status of high energy missions for the 1990s and beyond. There are 6 in orbit and 15 planned! The point is that there are other communities that are better at getting their missions into space than the infrared/submillimetre community. I would like to take a cue from our submillimetre brothers and sisters who a few months ago circulated through the international community a statement of outlook and principles asking for support of the submillimetre community for increased funding by the international and national space agencies for those long-term missions. I have enclosed a copy of the Manifesto – it simply states that there are certain important technologies that we think should be pursued and makes a variety of other points. Please read this and feel free to comment on it. Again thank you very much for your time.

LONGAIR – That really is an invitation for everyone to look at the Manifesto rather carefully and all constructive and non-constructive comments should go back to Harley.

LONGAIR – I'm going to make no summing up speech. I think the discussion we have had over the last couple of hours speaks for itself. Let me just make special tribute to both Tim and to Harley for their marvellous efforts. I think Tim deserves special mention because I remember the time when Tim first got the idea of passively cooled instruments a long time ago. Thank you both, gentlemen, for putting all this together.

MANIFESTO:

THE NEXT GENERATION INFRARED SPACE OBSERVATORY

1. The Importance of Infrared Astronomy

> The infrared and submillimeter portion of the spectrum ... is poorly explored but is of fundamental importance for almost all aspects of astronomy from solar system studies to cosmology.
>
> – US Astronomy and Astrophysics Survey Committee (1991)

Nature reveals many of its secrets to astronomers working at infrared wavelengths (1 μm to 100+ μm). Most of the line and continuum emission from solar system objects and planetary material around other stars emerges in this spectral region. Similarly, protostars, young stellar objects, and the molecular clouds out of which they form all emit the majority of their light in the infrared. Supernovae and novae, enriching the interstellar medium (ISM) with the elements necessary for life, have primary diagnostic lines at these wavelengths. Emission from nearby galaxies at shorter infrared wavelengths are dominated by their major stellar components, cool dwarf and giant stars, while at longer wavelengths we can observe prominent continuum and line emission from major phases of the interstellar material. Finally, at the greatest distances in the universe, expansion has redshifted major diagnostic continuum and line features into the infrared.

Telescopes on high mountain tops and in aircraft such as the Kuiper Airborne Observatory (KAO) and the planned Stratospheric Observatory for Infrared Astronomy (SOFIA) are powerful facilities that serve the international community of astronomers and play a critical rôle in developing new instrumentation. However, there are two fundamental limitations to ultimate infrared observations from within the earth's atmosphere. First, about half of the radiation from space between $\lambda = 1$ to 20 μm and virtually all of it beyond 30 μm is absorbed, primarily by water vapour. Second, even at relatively transparent wavelengths, such as those near 2 μm and 10 μm, the whisper of radiation from space is nearly overwhelmed by the intense foreground radiation from the atmosphere and warm telescope optics and support structure. Both problems have been solved by launching infrared telescopes into space and cooling them by surrounding them with massive Dewars filled with liquid helium.

2. Infrared Space Observatories

Only observations from space can obtain the extremely high sensitivities which are necessary to exploit the richness of the infrared. It has been the goal of major advances in our understanding of the cosmos via unimpeded infrared observations that has kept scientists, engineers, and administrators working for decades on past and current generations of space observatories. The international *Infrared Astronomical Satellite* (IRAS) and the *Cosmic Background Explorer* (COBE) demonstrated that nature cannot be understood without major exploration of these wavelengths. The European *Infrared Space Observatory* (ISO) will be launched soon and will extend our study of the infrared universe with a major increase in sophistication over IRAS. These first generation (cryogenic) missions will culminate with the *Space Infrared Telescope Facility* (SIRTF) with a further advance in performance over all earlier missions, which will draw to a close NASA's Great Observatories series.

These missions all used cryogen tanks to cool telescopes and detector elements. Although effective, evaporative cooling severely limits both telescope aperture and mission lifetime. Given the strong dependence of sensitivity upon aperture and the penalty of short space missions, second generation infrared space observatories must consider alternatives to this open-loop cooling. At present, the most promising is a mixture of radiative cooling augmented by closed-loop refrigerators, as proposed for *Edison*, a possible 2.5+ m space observatory now under preliminary study. However, despite its promise as a logical alternative to cryogenic cooling, radiative/mechanical systems face formidable technological hurdles.

Given that the lead time for some space missions can exceed a quarter century, a major effort must be undertaken now to both study plausible successors to the cryogenic missions and to control the administrative and budgetary processes that expand space missions to fill human lifetimes. With ISO only two years away from launch and SIRTF in advanced development, study of future technology is already overdue. If the length of time required to develop large space missions cannot be compressed, we can expect a second generation observatory to be launched no earlier than 2020. Once again, the time between major infrared missions will be measured in decades.

We conclude that any future infrared project, in addition to all other criteria for operation in space, must have as its goals [1] as large an aperture as is feasible, [2] an alternative to evaporative cooling so as to maximize longevity, [3] an international partnership, and [4] a substantially shorter lead time for design, development, and construction.

3. International Astronomy: Science without Frontiers

> ... [there should be] cooperation by the United States with other nations and groups of nations in work done pursuant to this act ...

– The US National Aeronautics and Space Act (1958)

> The committee recommends that international cooperation be considered for the development of a major initiative if such a project draws on complementary capabilities of different nations or requires resources beyond those that can be provided by the United States alone.

– US Astronomy and Astrophysics Survey Committee (1991)

Astronomy has no frontiers, either on earth or in the cosmos. International collaboration on scientific projects of all sizes is now commonplace and many nations have both the capability and desire to expand their operations in space. Furthermore, several countries possess the expertise to contribute toward solving the major technical hurdles identified for future space missions. It may be a combination of tradition and administrative inflexibility that is the most serious limitation to involving research teams from around the world.

At the same time, we must recognize legitimate national pride and interest in preserving technical and scientific resources. Just as a well-run team project will support individual goals and efforts, so too must a successful international space mission encourage the advancement of national scientific and technical expertise.

Nevertheless, constraining the minds and spirits of talented scientists, engineers, technicians, and administrators will most likely lengthen the development of new instrumentation for space and new solutions to difficult technical challenges. In the long run, restricting international collaboration may result in more costly missions. It is in the best interests of future projects to allow as many innovative minds and institutions to work on the very difficult problems of the next generation missions.

4. Technical Goals

> Technology development set in motion long before critical milestones is cost-effective. The committee concurs with the strong support ... for vigorous technology development across NASA's entire space astronomy programme.
>
> [Such technology development will] provide the basis for selection, *by the turn of the century*, of a new mission to follow the Great Observatories [emphasis added].

– US Astronomy and Astrophysics Survey Committee (1991)

Some technical goals are clear, based on several recent studies in several countries on post-cryogenic space observatories, including the Astronomy and Astrophysics Survey ('Bahcall') Committee report, Astrotech 21 workshops and technology assessments for the Next Generation Space Telescope (NGST) and *Edison* projects. Similar conclusions have been reached in other countries. As the assessments progress, the time will come to develop solutions to the major technological hurdles facing future infrared missions. With the enormous richness of the infrared, a powerful facility operating between a few and 100 μm must be among the post-Great Observatories that the Bahcall Committee expects to be considered within this decade.

At present, the most pressing challenges appear to be:

Space-qualified closed-cycle coolers. Very capable mechanical refrigerators are part of many scientific experiments on earth and closed-cycle coolers are being built for flight onboard the ESA ERS1 and NASA UARS satellites. However, the types of coolers likely to be required by non-cryogenic IR space missions are beyond current technology. Every effort must be expended to develop satisfactory non-cryogenic cooling in space.

Material performance at low temperatures. Despite more than a quarter century of experience in operation in cold space, there is significant ignorance about material performance at very low temperatures. This includes thermal properties of optical materials, performance of optical coatings as a function of temperature, thermal conductivity of support material at low temperature, and the combination of these effects (and many others) in a complete observatory system. Material performance at low temperatures must be a high priority goal in preparing for next generation missions.

Lightweight and active structures. A tremendous gain is realized by decreasing the mass of major components on a space mission. Similarly, long-lived active structures may be the preferred option to heavy 'dumb supports'. Promising advances in lightweight/active materials must be pursued.

Detector systems. Significant advances have been made in recent years in producing high-sensitivity, large-area detector arrays and their associated electronics and optics for space application. However, continued progress in such areas as sensitive operation at higher temperatures, radiation hardening, and near-chip signal processing is desirable for future space application. Support is necessary for further advances in detector systems technology for astronomical applications.

Satellite subsystems. Recent large space missions have demonstrated that major improvement in certain facility subsystems are extremely desirable for post-Great Observatory programs. These areas include onboard data manipulation, improved pointing/tracking control, and improved space-qualified computer systems. Resources must be devoted toward developing those future technologies that appear to have the best chance for significant improvement in spacecraft performance.

Design analysis of promising missions. Performance estimates for observatory/satellite subsystems require a systems analysis of the entire facility. Such analysis is necessary to determine, for example, power requirements, mass distribution, equilibrium temperatures, conductivity and rigidity requirements, and tolerances for optical designs. Early development of flexible designs must be undertaken to better define the requirements for next generation technology.

A NEXT GENERATION INFRARED SPACE OBSERVATORY

TOWARD THE 21st CENTURY

Given the major technological hurdles and long lead times in space missions, it is essential to begin *at once* a coordinated programme of study and technology development if a future generation infrared space mission will be a reality in the next decade. With our goal of construction of an international infrared space observatory within 10–20 years, we urge the national space agencies to pursue the following agenda.

- Continued unalloyed support for first generation infrared and sub-millimeter missions, including ISO, SIRTF, SOFIA, NICMOS, IRTS, and SWAS, as well as strong ground-based infrared astronomy programmes.
- Assessment of and support for promising technology likely to be necessary for future infrared (and other) space missions. This includes closed-cycle refrigerators, improved detectors, low temperature optics and coatings, lightweight and low-conductivity structures, and advanced satellite subsystems such as improved pointing controls and onboard data analysis.
- Support for advanced study of promising designs for future infrared space missions. This especially includes systems analysis of non-cryogenically-cooled space observatories.
- Support for technical and scientific studies broadly balanced among universities, research centres, aerospace corporations, and space agency facilities. Scientists must share rôles of leadership and responsibility in all cases.
- Support for strong international collaboration in future major infrared space missions at all levels. Such collaboration should be initiated, developed, and guided by scientists and technical personnel, supported by the national space agencies.

LIST OF PARTICIPANTS

D Y Anggraeni
c/o SRON-ROU
Sorbonnelaan 2
3589 CA Utrecht
The Netherlands

C Baddiley
DRA Electronics Division
St Andrew's Road
Malvern
Worcs
WR14 3PS
England

B Baud
Fokker
Postbus 12222
NL-1100 AA Amsterdam-Zuidoost
The Netherlands

P W J L Brand
University of Edinburgh
Department of Astronomy
Blackford Hill
Edinburgh EH9 3HJ
Scotland

S J Bell Burnell
Royal Observatory
Blackford Hill
Edinburgh EH9 3HJ
Scotland

S Chase
Instituto TESRE/CNR
Via de Castagnoli 1
I-40126 Bologna
Italy

J Curtis
Rutherford Appleton Laboratory
Chilton
Didcot
Oxon OX11 0QX
England

J K Davies
Royal Observatory
Blackford Hill
Edinburgh EH9 3HJ
Scotland

Jon Davies
University of Wales
Department of Physics
Park Place
Cardiff
Wales

R J Emery
Rutherford Appleton Laboratory
Chilton
Didcot
Oxon OX11 0QX
England

Th Encrenaz
DESPA
Observatoire de Paris-Meudon
92915 Meudon
Paris
France

E Erickson
NASA-Ames Research Center
MS245-6 Space Science Division
Astrophysics Exp. Branch
Moffett Field
CA 94035
USA

E Ettedgui-Atad
Royal Observatory
Blackford Hill
Edinburgh EH9 3HJ
Scotland

Space Science Reviews **61**: 265–268, 1992

J B Farrow
British Aerospace (Space Systems) Ltd.
Earth Observation & Science Division
FPC320
PO Box 5
Filton
Bristol BS12 7QW
England

S Fonti
Department of Physics
University of Lecce
73100 Lecce
Italy

I Furniss
Department of Physics & Astronomy
University College London
Gower Street
London WC1E 6BT
England

J W Harris
Royal Observatory
Blackford Hill
Edinburgh EH9 3HJ
Scotland

P M Harvey
Department of Astronomy
University of Texas
Austin, Texas 78712
USA

T G Hawarden
Joint Astronomy Centre
665 Komohana Street
Hilo, HI 96720
USA

M Hayashi
Department of Astronomy
University of Tokyo
Bunkyo-Ku
Tokyo 113
Japan

K A van der Hucht
SRON Space Research Laboratory
Sorbonnelaan 2
3589 CA Utrecht
The Netherlands

C M Humphries
Royal Observatory
Blackford Hill
Edinburgh EH9 3HJ
Scotland

P A James
Royal Observatory
Blackford Hill
Edinburgh EH9 3HJ
Scotland

B Jones
British Aerospace (Space Systems) Ltd.
Earth Observation & Science Division
FPC320, PO Box 5
Filton
Bristol BS12 7QW
England

M Kessler
ESA
Astrophysics Division
Keplerlaan 1
2201 AZ Noordwijk
The Netherlands

S Köse
Oerlikom Contraves AG
Schaffhauserstr 580
CH-8052
Zurich
Switzerland

T J Lee
Royal Observatory
Blackford Hill
Edinburgh EH9 3HJ
Scotland

M S Longair
University of Cambridge
Cavendish Laboratory
Madingley Road
Cambridge CB3 0HE
England

A J Longmore
Royal Observatory
Blackford Hill
Edinburgh EH9 3HJ
Scotland

A F J Morgovwnik
Fokker
Space & Systems
Postbus 12222
NL-1100 AE Amsterdam-Zuidoost
The Netherlands

C M Mountain
Royal Observatory
Blackford Hill
Edinburgh EH9 3HJ
Scotland

P G Murdin
Royal Observatory
Blackford Hill
Edinburgh EH9 3HJ
Scotland

V Natale
CNR-IROE
via Panciatichi 64
50127 Firenze
Italy

Y Nevo
Royal Observatory
Blackford Hill
Edinburgh EH9 3HJ
Scotland

A H Orlowska
Rutherford Appleton Laboratory
Chilton
Didcot
Oxon OX11 0QX
England

A Penny
Rutherford Appleton Laboratory
Chilton
Didcot
Oxon OX11 0QX
England

P Persi
Istituto di Astrofisica Spaziale
CNR
CP67 00044
Frascati
Italy

M Rowan-Robinson
Queen Mary & Westfield College
Mile End Road
London E1 4NS
England

S Rothero
OCLI Optical Coatings Ltd.
Ridge Way
Hillend Industrial Park
Dunfermline
Fife KY11 5FR
Scotland

S C Russell
Dublin Institute for Advanced Studies
School of Cosmic Physics
5 Merrion Square
Dublin 2
Eire

P Saraceno
CNR-IFSI
CP27 00044
Frascati (RM)
Italy

H van de Stadt
Space Research Groningen
Postbus 800
9700 AV Groningen
The Netherlands

R S Stobie
Royal Observatory
Blackford Hill
Edinburgh EH9 3HJ
Scotland

R I Thompson
Steward Observatory
University of Arizona
Tucson, AZ 85721
USA

H A Thronson
University of Wyoming
Department of Physics & Astronomy
PO Box 3905
University Station
Laramie, WY 82071
USA

M Ferrari-Toniolo
Istituto di Astrofisica Spaziale via E. Fermi 21
Box 67 00044
Frascati
Italy

R J Tuffs
Max Planck Institut für Kernphysik
Postfach 103980
D-6900
Heidelberg
Germany

H Walker
Rutherford Appleton Laboratory
Chilton
Didcot
Oxon OX11 0QX
England

R Waters
SRON Groningen
Laboratory for Space Research
Postbus 800
9700 AV Groningen
The Netherlands

M Wells
University of Edinburgh
Department of Astronomy
Blackford Hill
Edinburgh EH9 3HJ
Scotland

R D Wolstencroft
Royal Observatory
Blackford Hill
Edinburgh EH9 3HJ
Scotland

GUIDE AND EXAMPLES FOR USERS OF THE 'SPACEKAP' STYLE FILE

Basic Instructions

A. AUTHOR 1 and B. AUTHOR 2*
The Institute and its Address of authors 1 and 2.
and
C. AUTHOR 3
*Institute of Author 3***

Abstract. This document describes in some detail the use of the SPACEKAP style file for the preparation of articles to be submitted to journals or edited volumes published by Kluwer Academic Publishers. The 'spacekap' style file is very similar to the 'article' style sheet of LaTeX.

Key words: LaTeX – SPACEKAP – Guide

1. To Start

The 'SPACEKAP' style file is addressed by using the command: `\documentstyle{spacekap}`. The files SPACEKAP.STY and SPACE11.STY should be accessible from the directory you run LaTeX from.

2. The 'Opening' Environment

The title, subtitle, author names and institutes are handled by the *opening* environment in the preamble. Its structure is as follows:

```
\begin{opening}
\title{THE TITLE OF THE ARTICLE}
\subtitle{If There Is Any}
\author{Author's Name}
\institute{Institute and/or its Address}
\date{}
\end{opening}
```

2.1. Title and Subtitle

The title should be typed in CAPITALS. The subtitle in Initial Capitals and lowercase. The `\subtitle` command may be omitted. A missing `\title` command generates an error message.

* On leave from his original Institute
** affiliated to CMDA, Dordrecht

If your title or subtitle are too long for one line, you can split it by putting two backslashes at an appropriate place, e.g., the commands
`\title{GUIDE AND EXAMPLES FOR USERS OF \\ THE 'SPACEKAP' STYLE FILE}`
`\subtitle{Basic Instructions}`
have generated the title and subtitle of this Guide.

2.2. Author names and Institutes' Addresses

For each author's name the `\author` command should be used:
`\author{A. Author 1}`, `\author{B. Author 2}` etcetera. The names should be written in Initial Capitals and lowercase, but they are printed in Capitals.

The affiliation of authors is generated by the `\institute` command:
`\institute{The Institute and its Address}`
If there are more than one author at the same institute, you should first use the `\author` command for each author and then use the `\institute` command.

A second institute's name to indicate another affiliation can be handled through the `\thanks` command. For this Guide the following sequence of commands was used:

```
\author{A. Author 1}
\author{B. Author 2\thanks{On leave from his original institute}}
\institute{The Institute and its Address of \\authors 1 and 2.}
\author{C. Author 3}
\institute{Institute of Author 3\thanks {affiliated to CMDA,
Dordrecht}}
```

2.3. Date

The `\date` command gives you the possibility of entering the date (or a text) at the beginning of your article. If you do not wish a date, only the two brackets will do. In principle you can put any text between the brackets.

3. Running Heads

If the *title* of your article is longer than 60 characters, the running head will be omitted in the printout. You can make your own running title by using the `\runningtitle{}` command, where you put the shorter title in capitals between the curly brackets.

If the number of *authors* is more than two, please use the
`\runningauthor{}` command, and abbreviate the list of authors to `A. AUTHOR 1 ET AL.` which you put, in capitals, between the curly brackets (but ET AL. *not* in italics).

If you use one or both commands, please put these in the preamble, just before the `\begin{document}` command.

4. The Abstract Environment

This environment is standard LaTeX. You need not enter the word 'Abstract'. It prints the abstract in a smaller font than the body text.

4.1. Keywords

Key words are printed at the end of an abstract. The `\keywords` command should be used within the *abstarct* environment, but after the text of the abstract. For example, `\keywords \LaTeX -- SPACEKAP -- Guide` resulted in the key word line at the beginning of this Guide. Please, use `--` to separate the keywords.

5. Sections, Subsections and Subsubsections

A section starts with the command: `\section{The Title of This Section}`. If the (sub-)section title is too long, you can break it at an appropriate place by using `\protect\\`. (Only the two `\\`'s do not work here).

For the section, (sub)subsection titles you are kindly requested to use Initial Capitals and lowercase only, *not* entirely capitals.

6. Figures and Tables

6.1. Figures

The *figure* environment produces figure space and caption in the text. The numbering of figures is automatic.

```
\begin{figure}
\vspace{Space in mm or cm}
\caption{Caption Text.}
\end{figure}
```

The SPACEKAP style file puts figures wherever possible on the top of the page. This may not always work, in which case you need to be a little creative and consult the LaTeX manual.

6.1.1. Labeling of Figures

It may be convenient to label the figures in case you have to insert another figure later. The following can be done:

```
\begin{figure}
\vspace{}
\caption{}\label{short name}
\end{figure}
```

In the text you could write:
`... can be seen in Fig. \label{short name}.`

6.2. Tables

Tables can be generated in several environments for which we kindly like you to refer to the LaTeX manual.
An example of the convenient *tabular* environment for simple tables follows.

```
\begin{table}[h]
\caption{Caption text.}
\begin{center}
\begin{tabular}{lll}\hline
description 1 & description 2 & description 3\\
\hline\\
Entry1 & Entry 2 & Entry 3\\
Entry1a& Entry2a & Entry3a\\
\hline \end{tabular}
\end{center}
\end{table}
```

This produces:

TABLE I
Caption text.

description 1	description 2	description 3
Entry1	Entry 2	Entry 3
Entry1a	Entry2a	Entry3a

We wish to avoid both using vertical lines between columns and horizontal lines between rows.

6.2.1. Positioning of Tables

The position of a table in the text may not always be appropriate. As with figures, the style file tries to put it wherever possible on top of the page. In some occasions you may wish to put it at one particular spot in the text. The option `[h]` can then be used: `\begin{table}[h]`

It is acknowledged that positioning of figures and tables is by no means straightforward.

7. Acknowledgements

Acknowledgements can be given by using the `\acknowledgements` command. The text for this item can start on the same line as the command.

8. Appendices

An Appendix can be generated by using the `\appendix` command, followed by a `\section` command. This results in a Roman numbered Appendix section.

9. The References

References are inserted in the *thebibliography* environment:
`\begin{thebibliography}{}`. (The second pair of curly brackets is needed.)
For typing the references you may use `\bibitem{}`. In that case you must type the items as they should appear in the list:
`\bibitem{} Author, I.N.: year, 'Title of the Article', {\it journal}{\bf vol. no.}, page.`

9.1. (Author, Year) Referencing

For certain, rather standard, reference expressions there are a few commands available which generate the right output. Caution should be exercised in applying the correct number of items.

`\article{Author, A.N. and Secauth, N.}`
`{year}{journal abbrev.} {vol.no.}{page no.}` This command requires precisely *five* items. If an item is not known, the corresponding curly brackets should still be typed.

If the title of a reference is required or known, one can use the command:
`\reftitle {Author, A. and Secauth, N.}`
`{year} {title of article} {journal abbrev.} {vol.no.} {page no.}` This command requires precisely *six* items or {} combinations. If an item is not known, the corresponding curly brackets should still be typed.

`\book{Author, A.}{year}{title of book} {Publisher: city}{page no.}`
This command requires precisely *five* items. If an item is not known, the corresponding curly brackets should still be typed.

`\inproceedings{Author, A.} {year} {title of contribution} {title of proceedings} {editors} {Publisher:city} {page no.}`
This command is rather complex and needs some attention when typing the re-

quired *seven* entries. If an item is not known, the corresponding curly brackets should still be typed.

9.2. BIBTEX

You may use BIBTEX . Please be sure that you send all the necessary files with the original file, i.e. the `bbl` file.

10. Some Hints

A few typographical conventions should be adhered to in order to improve the readability of your text.

- As you will often be working in the mathmode, composing mathematical expressions, you will have to take action to prevent the subscripts from being printed in italics. For example, `$R_{max}$` produces: R_{max}. Convention is `$R_{\max}$` which gives: $R_{\max}$
- Do not use the mathmode to print italics, but use the `\it` command. $ITALICS$ in mathmode is different from *ITALICS* in italics.
- It is almost forbidden to use `\vspace` or `\hspace`. It often creates unexpected problems.
- Stacking of formulae must be avoided. The fontsizes often become too small to be read comfortably, or are even illegible.

11. Information

The '`spacekap`' style files are available from:

KLUWER ACADEMIC PUBLISHERS
P.O. Box 17, 3300 AA Dordrecht, The Netherlands
email: SCITECH@KAP.NL / fax: 31-78 334254

If you encounter problems, or when you have suggestions for improving the `spacekap` style file or this Guide, we would appreciate hearing from you.